T0176719

Shipboard Power Systems Design and Verification Fundamentals

Shipboard Power Systems Design and Verification Fundamentals

Mohammed M. Islam

Published by
Standards Information Network

WILEY

Registered Office
John Wiley & Sons, Inc., 111 River Street, Hoboken, NJ 07030, USA

Editorial Office
111 River Street, Hoboken, NJ 07030, USA

For details of our global editorial offices, customer services, and more information about Wiley products visit us at www.wiley.com.

Wiley also publishes its books in a variety of electronic formats and by print-on-demand. Some content that appears in standard print versions of this book may not be available in other formats.

Library of Congress Cataloging-in-Publication Data

Names: Islam, Mohammed M., author.
Title: Shipboard power systems design and verification fundamentals / Mohammed M. Islam.
Description: Hoboken, NJ : John Wiley & Sons, 2018. | Includes index. | Identifiers: LCCN 2018002475 (print) | LCCN 2018005907 (ebook) | ISBN 9781119084273 (pdf) | ISBN 9781119084143 (epub) | ISBN 9781118490006 (cloth)
Subjects: LCSH: Ships – Electric equipment – Design and construction. | Ships – Electronic equipment – Design and construction.
Classification: LCC VM471 (ebook) | LCC VM471 .I75 2018 (print) | DDC 623.87–dc23
LC record available at https://lccn.loc.gov/2018002475

Cover image: ©Dovapi/Gettyimages
Cover design by Wiley

Set in 10/12pt TimesLTStd by Aptara Inc., New Delhi, India

10 9 8 7 6 5 4 3 2 1

Dedicated to my wife, Raihana Islam

Contents

Preface

Shipboard electrical system design and development fundamentals have changed from traditional low-voltage to medium-voltage generation and distribution due to higher power requirements. Power electronics application is playing a major role, including adjustable speed propulsion drives and variable frequency drive for ship service auxiliary applications. The guidelines for shipboard use of medium-voltage adjustable speed drive (ASD) require further amplification. This book provides step-by-step details of widely accepted design applications for shipboard electrical engineering design fundamentals. These fundamentals are somewhat different for different class such as commercial ships, military ships, and offshore floating vessels. The design fundamentals of electrical power generation prime movers, the requirements of the distribution system, and the transition to various services, must meet safety requirements. These design and development fundamentals are presented to use as a guide for any new design. Additionally, design and verification of the design with multiple options is also a requirement, as modeling and simulation with hardware in the loop has become a norm at the fundamental design level. An attempt has been made to initiate design verification at a very early stage of design and carry it through to the detail design, procurement, installation, and commissioning stages.

The adjustable speed drive also contributes to major system-level electrical noise such as harmonics and transient instability. Harmonic requirements and guidelines such as IEEE 519 and IEEE 1584 play a vital role for industrial application. However, for the ungrounded shipboard power system one must be careful as to the use of IEEE 519 and IEEE 1584. The concept of complying with requirements for harmonic mitigation must be supported by the fact that the equipment will perform in a safe manner so that operators are safe, electrical power system coordination is properly engineered, and the transient aspect of the entire shipboard power system is managed very much within the required design boundary. Lack of verification of design fundamentals often leads to unsafe electrical systems. If the design is not properly integrated, at some point in the design and development phase additional corrective measures may be necessary to optimize it. However, the design solution may not be implemented due to practical constraints. At the preliminary or detail design phase, it is easy to demonstrate that the system will meet the requirement to enhance capability by using a system-level simulation such as a physics-based solution. A physics-based simulation such as Smart Ship System Design (S3D) is introduced to initiate an iterative process to prove the design concept with alternative choices and then select the one best suited for specific application.

The design and development of the shipboard power system is presented here as it began in the early 1970s; prior to that, the baseline shipboard power system design had been the same for many years. Electrical design and development changes accelerated when shipboard auxiliary systems of mechanical and hydraulic systems were being replaced with electrical systems. The author has gone through the real design challenge of developing an integrated electrical power system while designing a USCG Healy Icebreaker electrical propulsion medium-voltage distribution system. This involved the development of a medium-voltage system for shipboard adjustable speed propulsion for the Healy Icebreaker and medium-voltage generation and distribution for tankers. The design and development process was a challenge, such as, for example, to apply a 6-pulse versus a 12-pulse propulsion drive for the Healy Icebreaker and then quantify the total harmonic distortion (THD) of the electric propulsion system during the worst operational conditions.

The ship electrical system grounding requirement has become very challenging as the shipboard power system has taken a major shift from low-voltage ungrounded generation and distribution with a simple ground detection system to high-resistance grounding along with complex power generation and distribution requirements. This book addresses the medium-voltage distribution system with a resistance grounding system and then the impact of resistance grounding in view of ASD utilization-related grounding issues. The book provides multiple popular designs with resistance grounding, with variations to make designers aware of the implications of a concept which may or may not be considered an optimal design.

The shipboard electrical power system with high voltage and high power generation poses many challenging issues, such as complex, system-level protection coordination, sophisticated grounding requirements for ungrounded systems, special types of cable to deal with voltage surge and transients, harmonics, and special power filters for harmonic management.

At the system level of design and development, it has been recognized that a system with ASD may not maintain Class-I type power; the use of the Uninterruptable Power System (UPS) at higher power is being used. However, UPSs also bring solid state power electronic-related challenges.

The shipboard low voltage ungrounded power system ground detection and monitoring system is usually a simple detection system with lights for monitoring voltage variations. The IEC has developed completely different recommendations from the ground detection light with the understanding that the legacy system does not contribute to the management of real grounding danger, as the system leads to arcing and then bolted fault. The IEC requirement is to monitor and intervene as the electrical system starts making the transition from symmetric to asymmetric behavior. In case any ground is detected in the ungrounded electric system on ships, corrective action must be fast enough to protect the system from an arcing fault, explosion, and related equipment failures.

When a technological breakthrough challenges the real-life engineering application, sometimes failure may be encountered, which is the process of design and development. There must be a cause and effect analysis of the failure to get to the root cause and then take immediate corrective action. The corrective action process can be excruciating; however, finding a comprehensive and permanent solution is a must. Sometimes, multiple solutions may be adapted with multiple layers of protection to have a permanent solution. Whatever the design and development process is, the designer must have a thorough understanding of the solution being adopted.

This evolving engineering process of accepting technology's spiraling development is a normal developmental phenomenon. When a technology is accepted for development, it is considered to be working at present; however, it cannot be guaranteed for the future. As the technology is used, it becomes a candidate for standardization. Such is the case for IEEE 45-related standards for shipboard electrical power systems.

The selection of cable for an adjustable speed drive application is also a fundamental challenge of the IPS ship design application. This book provides in-depth analysis of cable-application challenges with recommended solutions.

As offshore-industry-related vessels embark on ASD-, VFD-, and AFE-type electrical installations, the challenging issues for the ships are also applicable.

The "all electric ship" concept of power generation and distribution with propulsion ASD and auxiliary system with VFD provide many operational advantages such as propeller torque delivery at any desired RPM and auxiliary system control at any speed. However, those controllers contribute other undesireable issues which the designer must understand and take appropriate measure. Some of these issues are:

- Electrical noise such as harmonics
- Understanding of VFD drive application, harmonic generation, harmonic calculation, harmonic management, special cable requirements, and special cable installation requirements
- Grounding matters at the generation level
- Grounding matters at the distribution level
- Grounding matters at the equipment level
- Single point grounding matters for MV and LV systems
- Medium voltage system protection and coordination
- Failure mode and its effect

The terminologies used for the design and development of VFD-related equipment mainly follow IEC standards. IEC terminologies and symbols are different from ANSI terminology and symbols. It is very important to understand the difference between IEC and ANSI standard electrical devices. These differences are identified along with examples, for the benefit of design engineers.

Grounding terminologies are different between ANSI and IEC standards. The IEC standard has PE, SG, and many other symbols associated with grounding. Those symbols create major confusion for design engineers.

This book provides guidelines emphasizing the safety and security of electrical and electronic equipment installation, equipment selection, and system coordination. The responsibility for implementing these recommendations belongs to everyone dealing with shipboard electrical equipment and electrical systems, such as electrical engineers, electrical designers, electrical cable pullers, electrical equipment installers, shipboard equipment and system testers, and troubleshooters.

At any voltage level, electricity is deadly. Traditionally, shipboard electrical voltage ratings have been 12 V, 24 V, 110 V, and 460 V for grounded and ungrounded installations. Until recently, the 460 V level was high for shipboard installation. In recent years the voltage level has risen to 4100 V, 6600 V, 11,000 V, and 13,800 V. The power requirement has increased from a few megawatts to hundreds of megawatts. Power generation and distribution at different voltages and at hundreds of megawatts have become a big challenge. IEEE Std 45 recommendations are a supplement to American Bureau of Shipping (ABS) rules and US Coast Guard (USCG) regulations for commercial ships. In the endeavor to standardize international rules and regulations, and with the advent of information technology, we have access to an enormous amount of technical information related to shipbuilding innovations, rules, regulations, and standards. Information technology has helped tremendously to make necessary information available at the click of a mouse. The responsibility to gain knowledge of available shipbuilding rules, regulations, and recommendations around the globe and adapt the most appropriate ones must be carried out at a very fast pace. The adaptation of the very process of technical innovation is also a universal challenge of building a bridge from present to future shipbuilding in order to meet tomorrow's demand.

The concept of IEEE Std 45 arose with the same objective as that of the National Electric Code® (NEC®). Acceptable standards are needed because no two persons will view something in the same way, interpret it in the same way, and implement it in the same way. These standards are critical in applying technology, which is a time-domain domino scenario by the very nature of innovation. As we build for the future, we have to live with the present. We must write down the most probabilistic aspect of an idea and agree to follow it. The accepted norm of today may not be the norm of tomorrow; however, it is appropriate today because it works to an accepted level and meets safety requirements.

Industry experts have contributed many years of experience in the shipboard electrical engineering field. Their task, however, has been presented with a significant challenge due to the global cooperation initiative, namely harmonization and globalization. IEEE Std 45 is in compliance with the NEC, the National Electrical Manufacturers' Association (NEMA), the Underwriters Laboratories (UL), the American Association of Testing and Material (ASTM), the American Bureau of Shipping (ABS) Rules, the Code of Federal Register (CFR) of the United States Department of Transportation, and various military specifications. The very process of equipment specification, manufacturing, installation, and testing has attained solid ground by the repeated revision of existing standards and the addition of new ones. IEC standards are also applicable for shipboard installation. The United States is a signatory to the IEC standards through the United States National Committee of the International Electrotechnical Commission, administered by the American National Standards Institute (ANSI). IEC standards differ from US standards in numerous ways, such as voltage level, unit of measurement, equipment rating, ambient rating, enclosure type, and equipment location classification. One must understand the differences to ensure applicability and interchangeability and combine the use of US standard equipment with IEC standard equipment. Most US standards committees have agreed to adopt IEC standards to supplement and change US standards. The IEEE Std 45 committee has also agreed to adopt IEC standards by directly replacing or modifying existing standards. These changes must be clearly understood in order to ensure that the safety and security of life and equipment are not compromised.

Smart Ship System Design (S3D) has been introduced as a new design environment with physics-based simulation and virtual prototyping of overall ship design, which is then compared with real system interaction for electrical power generation, distribution, protection, and automation.

There are many electrical one-line diagrams presented for design engineers who will be able to analyze different aspects of shipboard electrical distribution systems and then select the most appropriate one for application. If any one of the electrical one-line diagrams falls beyond the requirements of a regulatory body, the required correction must be made to ensure compliance.

This handbook is based on author's many years of ship building design experience and many years of experience in developing electrical standard for shipbuilding. The author wishes to thank all the individuals who have encouraged and contributed to the preparation of this book. The author also wishes to thank all IEEE 45 DOT standard working-group members for sharing technical know-how and expertise over the years, and technical experts in the marine field whose works may have been quoted in this handbook.

MOHAMMED (MONI) ISLAM

Chapter 1

Overview

1.0 INTRODUCTION

The shipboard electrical system design process consists of concept design, preliminary design, detail design, design development, design verification, installation, and commissioning. Shipboard power-system design and development is an engineering art that requires many years of engineering experience, specifically, designing electrical systems with experienced engineers. Shipboard electrical system design and development has become very challenging due to complex electrical power generation and distribution requirements including higher voltage, high power, and adjustable speed propulsion drives. The ship propulsion system has changed from direct mechanical drive to an electric motor with an adjustable speed drive. The across the line starters for auxiliary systems are being replaced with adjustable frequency/speed drive. Solid-state power electronics are being programmed to perform necessary ASD functions. However, solid-state devices and functionality also have some drawbacks, such as electrical noise. Most of the power electronic application-related hardware for shipboard application is migrated from well-established, shore-based industrial applications. There are subtle differences where industrial-based equipment is not suitable for shipboard applications. The shipboard power-system design process needs to be validated by methodical analysis with pros and cons. Sometimes the design process must go through a physics-based simulation process including hardware in the loop simulation to ensure that the design is optimized. The modeling and simulation of a shipboard electrical system provides many design options so that optimal design can be adapted for a custom shipboard design application. This book describes the following design and development process:

a. Basic design process, verification, and validation

b. Modeling and simulation-based design and verification

c. Smart ship system design (S3D)

Shipboard electrical power generation and distribution requirements are guided by rules, regulations, standards, and established recommendations by authorities

Shipboard Power Systems Design and Verification Fundamentals, First Edition. Mohammed M. Islam.
© 2018 the Institute of Electrical and Electronics Engineers, Inc. Published 2018 by John Wiley & Sons, Inc.

having jurisdiction in the design and development field. Power-system design engineers are to follow these guidelines to design required systems and get the design approved by the authority having approval jurisdiction. The shipboard low-voltage power system includes 1000 V, 690 V, 480 V, 230 V, and 120 V at 60 Hz and DC power at the voltage range from 12 V to 48 V, etc. The medium-voltage system includes all voltages from 1000 V to 35 kV AC as applicable for specific application. This book covers up to 15 kV maximum (11 kV or 13.8 kV nominal per MIL-STD-1399-300 and MIL-STD-680).

The shipboard power system consists of ship service power, emergency power, and propulsion power.

Shipboard power demand has evolved from a few megawatts to hundreds of megawatts. The voltage level has also been upgraded to 690 V, 2400 V, 4160 V, 6600 V, and beyond. Variable frequency drive or adjustable speed drive technology has become a dominant feature to mitigate propulsion-related higher voltage and high-power demand. The transition of proven VFD or ASD applications from industrial application to ship application has created many challenges.

The transition from low-voltage to medium-voltage generation and distribution may not have fathomed the requirements of grounded and ungrounded systems. The current practice of designing shipboard power generation and distribution systems may reflect a combination of both industrial and maritime applications.

Design engineers must understand the difference between industrial and ship applications of high voltage and high power. Design engineers must address these problems as uniform across all applications including shipbuilding; however, they should not arbitrarily consider the same solutions, as shipboard power generation and distribution fundamentals are different. For example, the harmonic noise problem can be addressed in general for all applications, but harmonic problem solution criteria for ships are different from those of other applications.

This handbook provides detail design and development of shipboard power generation and distribution based on low-voltage power generation and distribution, which has been well defined, as well as the medium-voltage system.

Model-based design has been presented to establish design variations and then the selection process starting from the concept design:

- Optimize the performance of the shipboard power system
- Optimize the functionality of the shipboard electrical network
- Define the requirements of electrical power equipment
- Define and coordinate the protective devices
- Reduce power losses
- Quantify harmonics contents and then systematically apply a harmonic management program to achieve an acceptable level
- Arc flash analysis and the establishment of a Hazard Risk Category (HRC)

1.1 SHIPBOARD POWER SYSTEM DESIGN FUNDAMENTALS

In general, a shipboard power system is ungrounded with delta distribution. The ground detection system is provided to detect ground in the system so that the ground is lifted as soon as possible. Single-phase ground is detectable, but will let the system continue to operate on the other two healthy phases. However, second ground fault, phase to phase will create arcing, which must be monitored and lifted as soon as possible. Three-system fault, which is also called bolted fault, must be detected as fast as possible and then the protective system must isolate the bolted fault to avoid any kind of explosion.

Power generation and distribution as well as solid-state devices operate with some ground reference. The basic requirements of a shipboard ungrounded system may be violated. The resistance grounding system using a wye-delta transformer with wye neutral connected to a ground with resistor also establishes a ground reference point in the ungrounded power system. The resistance grounded system and ungrounded low-voltage distribution system create a ground loop in the entire power system. The ungrounded power system ground plane in ideal conditions is a zero-voltage reference point. However, the combination of all ground paths in a shipboard power system may create the zero ground point other than zero plane to an unacceptable level.

The use of the delta-delta transformer has been changed over to ungrounded delta-wye or grounded delta-wye. The delta-wye configuration is not acceptable for shipboard installation as it can propagate electric noise with the wye distribution. The delta-delta is recommended as both primary and secondary will help to circulate electric noise within the winding. The delta-wye will circulate noise all over the distribution system due to the wye configuration. Again, it is very difficult to maintain zero ground reference in a wye distribution system. The grounded wye distribution system creates a ground plan coupled with an ungrounded zero reference point.

There are some special cases where a grounded wye distribution system is allowed due to operator safety reasons, such as an electrical workshop where the operator may use handheld electrical tools.

1.2 SHIP DESIGN REQUIREMENTS

Ship design USCG regulations, ABS rules, IEEE recommendations and IEC standards are used as appropriate. However, there are fundamental differences that must be taken under consideration. For example, the electrical grounding is "earthing" in the IEC standard. The three "A, B, C" phases are identified as "U, V, W," or "R, S, T," etc. These are very confusing, but the design and development expert must be familiar with them simply because IEC standards are equally recognized by US rules and regulations.

1.3 ETO CERTIFICATION: MEECE

The shipboard engine room watch-keeping Electro-Technical Officer (ETO) is required to go through training as evidence of competency. One of the competency training courses, "Management of Electrical and Electronic Control Equipment" (MEECE), is outlined in Chapter 13.

Each candidate for endorsement as an Electro-Technical Officer (ETO) on ships powered by main propulsion machinery of 750 kW/1,000 HP or more must provide evidence of having achieved the standard of competence specified in Table A-III/6 of the STCW Code (USCG 46 CFR 11.335(a)(2)). The table in this enclosure is adopted from Table A-III/6 of the STCW Code (found in Enclosure (4)) to assist the candidate and the assessor in the demonstration of competency.

1.4 LEGACY SYSTEM DESIGN DEVELOPMENT AND VERIFICATION

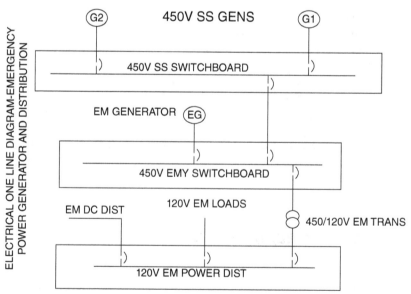

Figure 1.1   Typical EOL with Ship Service and Emergency Generator.

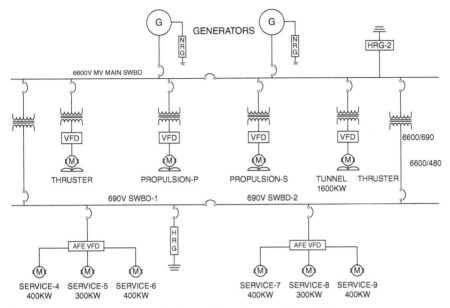

Figure 1.2   Typical EOL with Electric Propulsion and other VFD Services.

1.5 SHIPBOARD ELECTRICAL SYSTEM DESIGN VERIFICATION AND VALIDATION (V&V)

1.5.1 Verification and Validation (V&V) Overview

Shipboard electrical system design and development should have appropriate verification methods associated with them. Design verification must be traceable to a point that qualitative failure analysis (QFA) and design verification test procedure (DVTP) could be tied to each system as a deliverable.

The design verification and validation process must be traceable to operational scenarios so they are consistent with the Concept of Operations (CONOPS) of the ship.

1.5.2 Verification

Verification may be at the equipment level, system level, or system-of-system level. The design verification method should include the following:

i. Longhand calculation for preliminary design

ii. Modeling and simulation

iii. System-level engineering analysis

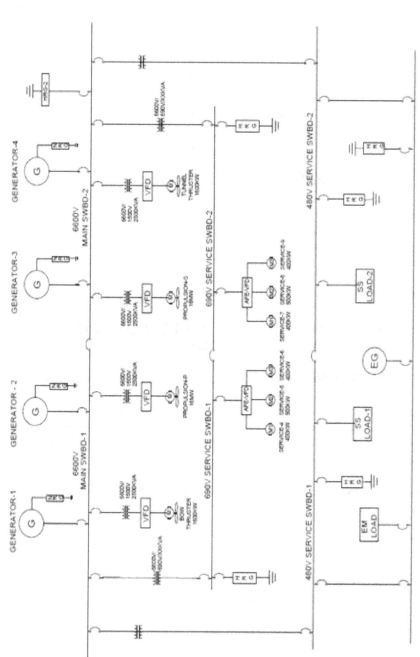

Figure 1.3 Typical EOL with All Electric Services and Emergency Generator.

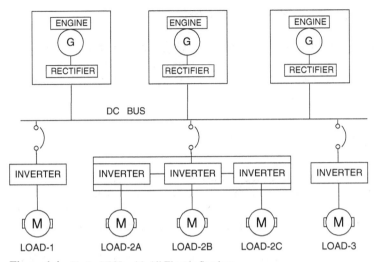

Figure 1.4   Typical EOL with All Electric Services.

AT THE RANGE OF 100MEGAWATT POWER
DISTRIBUTION

Figure 1.5   Typical Ship Variations for Commercial, USCG, Navy, and Offshore.

    iv. Complete system review

    v. Concept of operation development

    vi. Fundamentals QFA & DVTP

    vii. Compliance certification

1.5.2a Acceptance of Verification

Verification of the Support System involves verifying that each of the Support System Constituent Capabilities satisfy its relevant specification and that the Support System overall satisfies the requirements defined in the Support System Functional Baseline.

1.5.3 Validation

Validation of design should be conducted using scenarios that are consistent with the concept of operation. Because of the complexity of the systems that are being addressed and the significant time and effort required to conduct a comprehensive V&V program, the likelihood of completing a V&V program without the need for rework is low.

It is important that all test environments and equipment used during the V&V phases are controlled and validated to confirm that they will meet their objectives. If however the design development is verifiable as to the fact that the system has developed with a proven history of performance then it can be used, as prototype use may not require additional design verification.

1.5.4 Differences Between Verification and Validation: Shipboard Electrical System Design and Development Process

It is possible that an electrical design will pass the verification process but will fail when validated. It can happen that the design and development is in accordance with the specification, but the specification's shortcomings will lead to an overall non-functional ship or nonfunctional system. Therefore, as the design fundamentals are reviewed, proper verification and validation should be done to capture shortcomings and see appropriate corrective measures are taken.

The fundamentals of successful design are:

– regulatory body approval for the overall design
– individual equipment selection also with proven successful operation in the shipboard environment. Sometimes the equipment of a system works fine in the land-based installation but fails in the shipboard installation. Therefore it is mandatory to select marine duty equipment for shipboard installation.

The proliferation of VFD application has introduced many drawbacks and created many mishaps. Those drawbacks will be discussed for better understanding as to the overall design requirements and responsibilities.

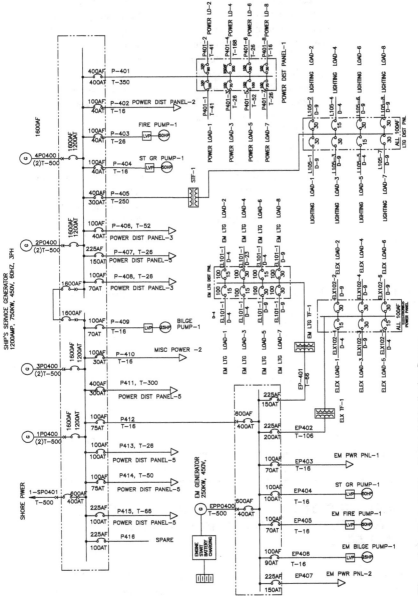

Figure 1.6 Typical EOL with Detail Distribution System.

Table 1.1   Typical Verification and Validation Table

Criteria	Verification	Validation
Definition	The process of evaluating electrical system design and development work deliverable products (not the actual final product procurement) of a development phase to determine whether they meet the specified requirements for the phase of development. The phases may be concept design, preliminary design or detail design.	The process of evaluating electrical system design development deliverables during or at the end of the development process to determine whether they satisfy the specified overall shipbuilding requirements.
Objective	To ensure that the design and development product is being developed (built) according to the requirements and design specifications. In other words, to ensure that work products meet their specified requirements.	To ensure that the product actually meets the customer's needs, and that the specifications were correct in the first place. In other words, to demonstrate that the product fulfills its intended use when placed in its intended environment.
Question	Are we designing and developing right for a functional ship?	Are we building the right ship?
Evaluation Items	Plans, Requirement Specs, Design Specs, Code, Test Cases	The actual functional ship.
Example-1	The requirement is to provide electric propulsion with 6-pulse drive. The drive selection verifies that the 6-pulse propulsion drive has been purchased and the system design has been developed accordingly.	The validation process has proved that though the procurement has met the customer requirement, it will not function as intended or will produce objectionable electrical noise which will contaminate the entire electrical system leading to a nonfunctional ship.
Example-2	Plans, Requirement Specs, Design Specs, Code, Test Cases	

1.6 IEEE 45 DOT STANDARDS: RECOMMENDED PRACTICE FOR SHIPBOARD ELECTRICAL INSTALLATION

The IEEE 45 standard development working group decided to further subgroup the standard with IEEE 45 DOT standards. This is to accommodate additional features of electrical installations on ships.

Therefore the IEEE 45 DOT standard subgroupings are:

IEEE 45 – Recommended Practice for Electrical Installations on Shipboard: Base Document (under development)

IEEE 45.1 – Shipboard Design and Development

IEEE 45.2 – Shipboard Controls and Automation

IEEE 45.3 – Shipboard System Engineering

IEEE 45.4 – Marine Sectors and Functions

IEEE 45.5 – Shipboard Electrical Safety Considerations

IEEE 45.6 – Shipboard Electrical Testing

IEEE 45.7 – Shipboard AC Switchboards

IEEE 45.8 – Shipboard Cable Installations

1.7 OTHER RULES AND REGULATIONS, AND STANDARDS IN SUPPORT OF IEEE 45 DOT STANDARDS

NEC – National Electrical Code

NFPA 70-E

IEEE 1662

IEEE 1580 – Standard for Shipboard Cable Construction

IEEE 1580.1 – Standard for Shipboard Bus-Pipe Installations

IEEE/IEC/ISO 80005.1 – Standard for MV Ship-to-Shore Power System

1.8 SHIPBOARD UNGROUNDED POWER SYSTEM

UNGROUNDED 3 WIRE   DELTA POWER GENERATION AND DISTRIBUTION SYSTEM (RECOMMENDED FOR SHIPBOARD USE. MOSTLY COVERED IN THIS BOOK)

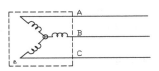

UNGROUNDED   WYE, 3 WIRE, SHIPBOARD POWER GENERATION AND DISTRIBUTION SYSTEM (RECOMMENDED FOR SHIPBOARD USE. MOSTLY COVERED IN THIS BOOK)

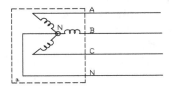

GROUNDED 4 WIRE WYE SYSTEM. NOT RECOMMENDED FOR SHIPBOARD POWER GENERATION AND DISTRIBUTION SYSTEM. LIMITED DISCUSSION IN THIS BOOK TO COMPARE BETWEEN UNGROUNDED AND GROUNDED SYSTEM

Figure 1.7   Typical 3-4 Wire Distribution System.

1.9 SHIPBOARD ELECTRICAL DESIGN BASICS

This book covers electrical power system detailed design and development for commercial ships such as cruise ships, cargo ships, tankers, related support vessels, offshore industry-related floating platforms, and all other support vessels. Some military vessel ship designs are also included to establish basic design fundamental differences as to redundancy requirements and zonal distributions. These design requirements and fundamentals are with the understanding of the following:

- Regulatory requirements

- Operational requirements

- Redundancy requirements

- Understanding of emergency requirements as to power generation as well as emergency load distribution

- Understanding the causes of a blackout (dead ship) situation. The blackout situation for all electric ship-related power generation and distribution is more complex than for the ship with nonelectric propulsion.

- Electric propulsion-related power generation and distribution requirements have been taken to adapt medium voltage power generation, due to the fact that ample power is available to change a hydraulic system or mechanical system to an electric system with variable drive operation

- The grounding requirements are different than the traditional low-voltage distribution though both systems are three-wire ungrounded systems

- A vital auxiliary must be properly classified as one design may be different than the other due to operational requirements

- There are regulatory requirements of vital auxiliary-related redundant services and operational requirements that directly contribute to the design and development

In general, a shipboard electrical system is ungrounded with few exceptions. The ungrounded system is only there is no dedicated neutral line in the distribution system. However, there always exists a capacitive ground path. This phenomenon needs to be explained with grounding and bonding. For better understanding, grounding and bonding will be called "G." Otherwise, the neutral line will be called "N." Nonlinear solid-state power applications usually create rapid changes to voltage and current while transferring energy to the load. These changes cause high-frequency current to flow to the ground. This is considered electrical noise.

There are many good features of electric drive-related applications onboard ships and platforms. However, many features may contribute electric noise, such as harmonics, transients, and grounding at the equipment level and system level. The design engineer must understand those issues so that causes and effects are properly analyzed during concept design and detail design. Recent VFD-related failure reports warrant better understanding, better design, and then overall design management.

Electrical propulsion and auxiliary service requirements for the use of variable frequency drive have contributed to recent operational challenges due to critical operational issues.

Regulatory bodies are in the process of addressing some of those issues and are considering additional regulations. The International Maritime Organization (IMO), under the Safety of Life at Sea (SOLAS), is considering additional redundancy considerations for critical propulsion-related auxiliary redundancy and further consideration of emergency power management. The requirements of these new initiatives are beyond the scope of this book; however, some will be addressed as a specific design is presented. These additional requirements will be addressed along with the proven requirements for better understanding of the origins of those requirements, and then added to the design.

There are many sample designs presented in this book, to show that each and every operational requirement is unique, leading to a customized design. The designs are presented with drawings and diagrams, so fundamental steps are discussed and then compared with the service requirements. This includes ship service power requirements, power requirements, and emergency power requirements. The design developments are presented strictly in compliance with regulatory body requirements.

The design includes:

(1) Shipboard electrical low-voltage and medium-voltage power generation, electrical propulsion, and power distribution systems. The fundamental shipboard electrical design requirements, design details, verification of the design prior to equipment installation, and then verification of the test results to establish a design base for the ship.

(2) Offshore floating platforms and offshore support vessels as applicable.

The variable frequency drive application has become an integral part of the electrical propulsion system, thruster applications, and other ship service auxiliary loads. High power requirements such as a propulsion system require high voltage and power. The power conversion of fundamental AC power to DC and then to AC conversion comes with many different technologies and challenges. Some of those challenges are at the electrical-system level and some are at the VFD-equipment level. These technical issues and challenges are presented for a basic understanding of the fundamentals as well as design guidelines so that design engineers can use this book as a reference. Details of practical design examples are also presented so that engineering students can learn the fundamentals of electrical power system design and development.

The shipboard electrical power system is usually ungrounded, with some exceptions. All design and development will be analyzed for ungrounded systems only. Different system-level groundings are presented with examples and recommendations so that design engineers understand the complexity of the grounding issues and then decide on a specific type of grounding as applicable and best suited for the installation.

Due to high power requirements for many applications such as propulsion-related services, medium-voltage generation and distribution is a requirement. This leads to many new technologies such as VFD, power filtering, EMC filters, dynamic braking systems, etc.

Due to the use of national and international standards for design and equipment selection, design engineers are challenged with many issues where compromises must be made. There are cases where compromise may not be in the best interest of the overall system design as well as the selection of equipment. It has been decided to identify the issues related to the process of harmonization of different standards with limited understanding of the consequences. Those issues are identified as much as possible so that the decision process can be better informed.

All examples are presented in this book mostly in view of practical experience and installation in compliance with rules and regulations for the intended operation, and required protection for the safety of the equipment, system, and operators.

1.10 ELECTRICAL DESIGN PLAN SUBMITTAL REQUIREMENTS

USCG Code of Federal Regulations CFR 46 Part 110

Subpart 110.25—Plan Submittal

§110.25-1 Plans and information required for new construction.   The following plans, if applicable to the particular vessel, must be submitted for Coast Guard review in accordance with §110.25-3:

> (a) Elementary one-line wiring diagram of the power system, supported by cable lists, panel board summaries, and other information including—
> (1) Type and size of generators and prime movers;
> (2) Type and size of generator cables, bus-tie cables, feeders, and branch circuit cables;
> (3) Power, lighting, and interior communication panel boards with number of circuits and rating of energy consuming devices;
> (4) Type and capacity of storage batteries;
> (5) Rating of circuit breakers and switches, interrupting capacity of circuit breakers, and rating or setting of overcurrent devices;
> (6) Computations of short circuit currents in accordance with Subpart 111.52; and
> (7) Overcurrent protective device coordination analysis for each generator distribution system of 1500 kilowatts or more that includes selectivity and shows that each overcurrent device has an interrupting capacity sufficient to interrupt the maximum asymmetrical short-circuit current available at the point of application.
>
> (b) Electrical plant load analysis including connected loads and computed operating loads for each condition of operation.
>
> (c) Elementary and isometric or deck wiring plans, including the location of each cable splice, a list of symbols, and the manufacturer's name and identification of each item of electrical equipment, for each—

    (1) Steering gear circuit and steering motor controller;
    (2) General emergency alarm system;
    (3) Sound-powered telephone or other fixed communication system;
    (4) Power-operated boat winch;
    (5) Fire detecting and alarm system;
    (6) Smoke detecting system;
    (7) Electric watertight door system;
    (8) Fire door holding systems;
    (9) Public address system;
    (10) Manual alarm system; and
    (11) Supervised patrol system.

(d) Deck wiring or schematic plans of power systems and lighting systems, including symbol lists, with manufacturer's name and identification of each item of electric equipment, and showing:
    (1) Locations of cables;
    (2) Cable sizes and types;
    (3) Locations of each item of electric equipment;
    (4) Locations of cable splices.

(e) Switchboard wiring diagram.

(f) Switchboard material and nameplate list.

(g) Elementary wiring diagram of metering and automatic switchgear.

(h) Description of operation of propulsion control and bus transfer switchgear.

1.11 ABS RULES FOR BUILDING AND CLASSING STEEL VESSELS

ABS-SVR: Part 4, Chapter 8, Section 1

V Electrical Systems: General Provisions 4-8-1

ABS5 Plans and Data to Be Submitted

One-Line Diagram:

One-line diagram of main and emergency power distribution systems to show:

    A. Generators: kW rating, voltage, rated current, frequency, number of phases, power factor.

    B. Motors: kW or HP rating, voltage, and current rating.

    C. Motor controllers: type (direct-on-line, star-delta, etc.), disconnect devices, overload and undervoltage protections, and remote stops, as applicable.

    D. Transformers: kVA rating, rated voltage and current, winding connection.

    E. Circuits: designations, types and sizes of cables, trip setting and rating of circuit protective devices, rated load of each branch circuit, emergency tripping and preferential tripping features.

F. Batteries: type, voltage, rated capacity, conductor protection, charging and
discharging boards.

ABS 5.1.2 Schematic Diagrams

Schematic diagrams for the following systems are to be submitted. Each circuit
in the diagrams is to indicate type and size of cable, trip setting and rating of circuit
protective device, and rated capacity of the connected load.

General lighting, normal and emergency

Navigation lights

Interior communications

General emergency alarm

Intrinsically safe systems

Emergency generator starting

Steering gear system

Fire detection and alarm system

ABS 5.1.3 Short-Circuit Data

Maximum calculated short-circuit current values, both symmetrical and asymmetrical values, available at the main and emergency switchboards and the
down stream distribution boards.

Rated breaking and making capacities of the protective devices. Reference may
be made to IEC Publication 61363-1 Electrical Installations of Ships and
Mobile and Fixed Offshore Units – Part 1: Procedures for Calculating Short-
Circuit Currents in Three-Phase A.C.

ABS 5.1.4 Protective Device Coordination Study

This is to be an organized time-current study of all protective devices, taken in
series, from the utilization equipment to the source, under various conditions
of short circuit. The time-current study is to indicate the settings of long-
time delay tripping, short-time delay tripping, and instantaneous tripping, as
applicable. Where an overcurrent relay is provided in series and adjacent to
the circuit protective devices, the operating and time-current characteristics of
the relay are to be considered for coordination. Typical thermal withstanding
capacity curves of the generators are to be included, as appropriate.

ABS 5.1.5 Load Analysis

An electric-plant (including high-voltage ship service transformers or converters,
where applicable per 4-8-2/3.7) load analysis is to cover all operating conditions of
the vessel, such as conditions in normal seagoing, cargo handling, harbor maneuver,
and emergency operations.

ABS 5.1.6 Other Information

– Description of the power management system, including equipment fitted with
preferential trips.

– Schedule of sequential start of motors, etc., as applicable.

– Voltage-drop for the longest run of cable of each size.

– Maintenance schedule of batteries for essential and emergency services.

– Plans showing details and arrangements of oil mist detection/monitoring and alarm arrangements.

– Information on alarms and safeguards for emergency diesel engines.

1.12 SHIPBOARD ELECTRICAL SAFETY CONSIDERATIONS

In general, shipboard electrical safety considerations recommendations are as in IEEE 45.5, NFPA 70E and other applicable standards. The safety considerations are outlined as:

"All electrical facilities shall be installed and maintained in a safe manner. All work involving electrical energy must be performed in a safe manner. The primary safe work practice is to establish an electrically safe work condition. The policy of this shipboard electrical design, development and operation is to implement the requirements found in NFPA 70E, Standard for Electrical Safety in the Workplace."

A basic rule that should be derived from the policy statement is that work on or near any exposed energized electrical conductors or circuit parts should be prohibited, except under justified, controlled, and approved circumstances, knowing that exceptions to this policy may become necessary. For example, measuring voltage or current are common and necessary tasks that involve exposure to energized conductors. Guidelines for such justification are provided in NFPA 70E, Section 130.2: *"Energized electrical conductors and circuit parts to which an employee might be exposed shall be put into an electrically safe work condition before an employee performs work."* NFPA Section 130.2(A) goes on to provide the acceptable exceptions to this rule.

Under blackout situations there are specific requirements to restore power. Some requirements are with specific time ranges. This is an attempt to identify these requirements and analyze those requirements. The requirements are given as follow to better indicate how to comply with those requirements.

– Starting and connection to the main switchboard of the standby generator are to be preferably within 30 seconds, but in any case not more than 45 seconds, after loss of power.

– Where electrical power is necessary to restore propulsion, the capacity shall be sufficient to restore propulsion to the ship in conjunction with other machinery, as appropriate, from a dead ship condition within 30 minutes after blackout.

– The emergency generator and other means needed to restore propulsion are to have a capacity such that the necessary propulsion-starting energy is available within 30 minutes of the blackout/dead ship condition as previously defined.

Emergency generator stored starting energy is not to be directly used for starting the propulsion plant, the main source of electrical power and/or other essential auxiliaries (emergency generator excluded).

– Emergency generators should be capable of carrying a full rated load within 45 seconds after the loss of the normal power source.

– Take home power requirements:

All examples are presented in this book mostly in view of practical experience and installation in compliance with rules and regulations for the intended operation, and the required protection for the safety of equipment, system and operators.

The ship steering gear and maneuvering system is considered one of the most critical systems on board ship.

– In case of main power loss to the steering gear, power must be totally restored within 45 seconds. This can be accomplished by the automatic bus transfer feature of the emergency switchboard. In case of a blackout of ship service power, there must be an automatic start-up of the emergency generator and transfer to emergency power.

– The alternate power must be capable of continuously operating for half an hour for steering the ship from 15 degree to 45 degree roll to either side in not more than 60 seconds, at maximum design draft loading, and half of the maximum design speed or 7 knots, whichever is greater.

1.13 HIGH-RESISTANCE GROUNDING REQUIREMENTS FOR SHIPBOARD UNGROUNDED SYSTEMS (SEE CHAPTER 9 FOR DETAILS)

(a) Background of High-Resistance Grounding (HRG) with Maximum Fault Current 5 Amp RMS

In 1935, a 40 ohm neutral to ground resistor was installed as a "safety resistor" on a 2300 V (L-N) minning shovel feed transformer to limit ground fault current to 55 Amp, a value sensed by protective relays of the era. This resistor, along with a guaranteed Ground Grid resistance of 2 ohms, limited line-ground fault voltage on the frame to a maximum 110 V potential, considered a "safe" level. It was also duly noted that single line-fault damage to cable, motors, and other electrical equipment was greatly reduced. Resistance grounded distribution systems found widespread use in the 1940s and 1950 s for industrial application with continuous process where a first line-ground fault could not be tolerated or would result in a more hazardous situation than automatic circuit breaker trip-off. A common problem was the lack of a fault location technology to quickly find the fault that would appear decades later. For comparison, modern low-voltage high-resistance grounding systems usually have a maximum fault current of 2 to 5 Amps.

1.14 SHIPBOARD ELECTRICAL SAFETY CONSIDERATIONS

1.14.1 Arc Flash Basics (See Section 12 for Details)

An arc flash is a dangerous condition associated with the release of energy caused by an electric arc, as per *IEEE 1584 Guide for Performing Arc Flash Hazard Calculations.*

Over the past two decades the electrical industry has begun to recognize arc flash as a serious safety hazard for anyone working near exposed, energized conductors and equipment.

NFPA 70E: Standard for Electrical Safety in the Workplace, requires that equipment be clearly labeled with safety information to minimize risk of injury to personnel. The National Electrical Code (NEC) article 110-16 requires arc flash hazard warning labels.

In addition, OSHA 29 CFR-1910 Subpart S regulates and states in part that safety practices shall be employed to prevent electric shock or other injuries resulting from either direct or indirect electrical contacts.

Electrical equipment, such as switchboards, panel boards, industrial control panels, meter socket enclosures, and motor control centers, that are in other than dwelling units, and are likely to require examination, adjustment, servicing, or maintenance while energized shall be field marked to warn qualified persons of potential electric arc flash hazards. See Figure 1 for an example of arc flash study results.

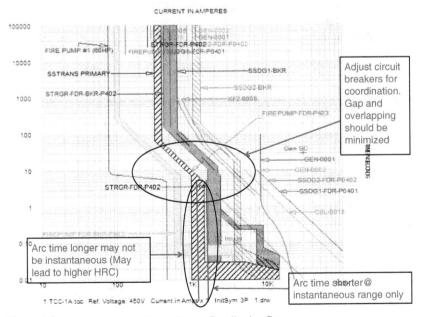

Figure 1.8 Typical Shipboard Power System Coordination Curves.

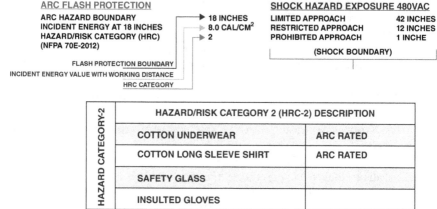

Figure 1.9 Typical Shipboard Power System ARC Flash Label.

At the completion of the study, 4" × 6" vinyl labels meeting the requirements of ANSI Z535 are furnished. The labels indicate the information required per NFPA 70E including the arc flash hazard boundary and the PPE (Personal Protective Equipment) levels. The labels are white vinyl, suitable for indoor or outdoor use, and printed on a thermal imaging printer. An example of an arc flash label is indicated in Figure 1.9.

1.14.2 Arc Flash Hazard Analysis Procedures

To conduct an arch flash hazard analysis, we go through the following steps per IEEE 1584:

1. Collect system and installation data
2. Determine the system modes of operation
3. Determine the bolted fault currents
4. Determine the arc fault currents
5. Find the protective device characteristics and the duration of the arcs
6. Document the system voltages and classes of equipment
7. Select the working distance

8. Determine the incident energy for all equipment

9. Determine the flash protection boundary for all equipment

1.14.3 Warning Label Placement

Labels shall be located so as to be clearly visible to qualified persons before examination, adjustment, servicing, or maintenance of the equipment.

The following equipment is included in an arc flash study and will be furnished with warning labels:

- 15 kV metal clad switchgear
- 5 kV metal clad switchgear
- 5 kV motor controllers
- 480 V low-voltage drawout switchgear
- 480 V motor control centers
- 480 V panelboards
- 480 V disconnect switches
- 208 V equipment served from a transformer greater than 125 kVA

1.15 PROPULSION POWER REQUIREMENTS (IEEE STD 45-2002, CLAUSE 7.4.2)

In determining the number and capacities of generating sets to be provided for a vessel, careful consideration should be given to the normal and maximum load demands (i.e., load analysis) as well as the safe and efficient operation of the vessel when at sea and in port. The vessel must have at least two generating sources. For ships, the number and ratings of the main generating sets should be sufficient to provide one spare generating set (one set not in operation) at all times to service the essential and habitable loads.

For MODUs, with the largest generator offline, the combined capacity of the remaining generators must be sufficient to provide normal (non-drilling) load demands.

For vessels propelled by electric power and having two or more constant-voltage, constant-frequency, main power generators, the ship service electric power may be derived from this source and additional ship service generators need not be installed, provided that with one main power generator out of service, a speed of 7 knots or one-half of the design speed (whichever is the lesser) can be maintained. The combined normal capacity of the operating generating sets should be at least equal to the maximum peak load at sea. If the peak load and its duration are within the limits of the specified overload capacity of the generating sets, it is not necessary to have the combined normal capacity equal to the maximum peak load.

1.16 IMO-SOLAS ELECTRIC PROPULSION POWER REDUNDANCY REQUIREMENTS

(a) Electrical Propulsion and Blackout

ELECTRIC PROPULSION REQUIREMENT: IMO REQUIREMENTS

Main Source of Electrical Power

(IMO-SOLAS Regulation II-1/41.5)

Interpretation of the clause "will be maintained or immediately restored" as detailed in Reg. II-1/41.5.1.1 amending SOLAS Reg. II-1/41 - Main Source of electrical power and lighting systems.

1. Reg. II-1/41.5.1.1 - Where the main source of electrical power is necessary for propulsion and steering of the ship, the system shall be so arranged that the electrical supply to equipment necessary for propulsion and steering and to ensure safety of the ship will be maintained or immediately restored in case of loss of any one of the generators in service.

2. To fulfill the above the following measures are required:

   2.1 Where the electrical power is normally supplied by more than one generator set simultaneously in parallel operation, provision of protection, including automatic disconnection of sufficient non-essential services and if necessary secondary essential services and those provided for habitability, should be made to ensure that, in case of loss of any of these generating sets, the remaining ones are kept in operation to permit propulsion and steering and to ensure safety.

   2.2 Where the electrical power is normally supplied by one generator, provision shall be made, upon loss of power, for automatic starting and connecting to the main switchboard of stand-by generator(s) of sufficient capacity with automatic restarting of the essential auxiliaries, in sequential operation if required. Starting and connection to the main switchboard of the stand-by generator is to be preferably within 30 seconds, but in any case not more than 45 seconds, after loss of power.

   Where prime movers with longer starting time are used, this starting and connection time may be exceeded upon approval from the society.

   2.3 Load shedding or other equivalent arrangements should be provided to protect the generators required by this regulation against sustained overload.

   2.3.1 The load shedding should be automatic.

   2.3.2 The non-essential services, service for habitable conditions may be shed and where necessary, additionally the secondary essential services, sufficient to ensure the connected generator set(s) is/are not overloaded.

(b) IMO-SOLAS Emergency Requirements

IMO-SOLAS Emergency Source of Power in Passenger and Cargo Ships

IMO-SOLAS Reg. II-1/42.3.4 and II-1/43.3.4

SOLAS Regulations II-1/42 and II-1/43 address emergency sources of electrical power in passenger ships and cargo ships respectively. Regulations II-1/42.3.4 and II-1/43.3.4 read as follows:

For ships constructed on or after 1 July 1998, where electrical power is necessary to restore propulsion, the capacity shall be sufficient to restore propulsion to the ship in conjunction with other machinery, as appropriate, from a dead ship condition within 30 minutes after blackout.

Interpretation:

"Blackout" as used in Regulation II-1/42.3.4 and II-1/43.3.4 is to be understood to mean a "dead ship" condition.
   "Dead ship" condition, for the purpose of Regulation II-1/42.3.4 and II-1/43.3.4, is to be understood to mean a condition under which the main propulsion plant, boilers, and auxiliaries are not in operation and in restoring the propulsion, no stored energy for starting the propulsion plant, the main source of electrical power and other essential auxiliaries is to be assumed available. It is assumed that means are available to start the emergency generator at all times.
   The emergency generator and other means needed to restore the propulsion are to have a capacity such that the necessary propulsion starting energy is available within 30 minutes of blackout/dead ship condition as defined above. Emergency generator stored starting energy is not to be directly used for starting the propulsion plant, the main source of electrical power and/or other essential auxiliaries (emergency generator excluded).

1.17 REGULATORY REQUIREMENTS FOR EMERGENCY GENERATOR

There are strict requirements as to the size selection, space selection and dead ship starting requirements for the safety of life at sea. For detailed requirements refer to IMO-SOLAS regulations, USCG regulations, ABS rules and IEEE-45 recommendations. Special consideration for the use of an emergency generator:

ABS-SVR-4-8-2-5.17 Use of Emergency Generator in Port (2002)
   Unless instructed otherwise by the Flag Administration, the emergency generator may be used during lay time in port for supplying power to the vessel, provided the following requirements are complied with. Arrangements for the Prime Mover

   (a) Fuel oil tank. The fuel oil tank for the prime mover is to be appropriately sized and provided with a level alarm, which is to be set to alarm at a level where there is still sufficient fuel oil capacity for the emergency services for the period of time required by ABS-SVR 4-8-2/5.5.

(b) Rating. The prime mover is to be rated for continuous service.

(c) Filters. The prime mover is to be fitted with fuel oil and lubricating oil filters in accordance with ABS-SVR 4-6-5/3.5.4 and 4-6-5/5.5.2, respectively.

(d) Monitoring. The prime mover is to be fitted with alarms, displays and automatic shutdown arrangements as required in 4-9-4/Table 8, except that for fuel oil tank low-level alarm, ABS-SVR 4-8-2/5.17.1(a) above is to apply instead. The displays and alarms are to be provided in the centralized control station. Monitoring at the engineers' quarters is to be provided as in ABS-SVR 4-9-4/19.

United States Coast Guard regulations make no mention of the use of an emergency generator in port as outlined in the ABS requirements. Due to a deepwater horizon incident, the USCG issued directives for DP vessels in 2012 as a draft, which is in full effect. Therefore, USCG regulations must be complied with for all new building.

1.18   USCG DYNAMIC POSITIONING (DP) GUIDELINES

(a) Background

Over the past several decades, the expansion of offshore exploration, development and production into deeper water has transformed an industry once characterized by relatively simple, domestic shallow water fixed platforms and small logistical vessels into an industry with complex, international floating vessels supplied and serviced by other large, international multipurpose vessels. This has given rise to the use of DP as a practical means for keeping these vessels within precise geographic limits. Failure of a DP system on a vessel conducting critical operations such as oil exploration and production could have severe consequences including loss of life, pollution, and property damage. This is particularly true for Mobile Offshore Drilling Units (MODUs), where a loss of position could result in a subsea spill and potentially catastrophic environmental consequences. The Deepwater Horizon incident demonstrated the serious challenges associated with subsea spill response. In a preliminary effort to better understand critical systems, training, and emergency procedures put in place to prevent or mitigate a loss of position on a dynamically positioned MODU and inform any related future rulemaking, the Coast Guard published a notice in the Federal Register (76 FR81957) requesting public comment on a draft policy. We received comments both as submissions to the docket and at a public meeting held on February 9, 2012. The Coast Guard was encouraged to publish a rule for areas where no standard has been set and to consider industry standards and guidance when developing the rule. The Coast Guard agrees and intends to initiate a rulemaking that addresses DP incident reporting requirements and minimum DP system design and operating standards

(b) USCG Interim Voluntary DP System Guidance

On July 7th, 2010, in response to a request from the Coast Guard, NOSAC issued the report "Recommendations for Dynamic Positioning System Design and Engineering, Operational and Training Standards." The report contained draft guidelines from the Marine Technology Society (MTS) Dynamic Positioning Committee, which the MTS has since completed. The Coast Guard has reviewed the guidance, referred to it when responding to known DP incidents and found it to be comprehensive and highly useful. Until the Coast Guard publishes a DP Rule, the Coast Guard recommends owners and operators of dynamically positioned MODUs (not leaseholders who contract MODUs) operating on the U.S. Outer Continental Shelf (OCS) voluntarily follow guidance provided in the "DP Operations Guidance Prepared through the Dynamic Positioning Committee of the Marine Technology Society to aid in the safe and effective management of DP Operations," March 2012 Part 2 Appendix. It is particularly important they identify the DP System's Critical Activity Mode of Operation (CAMO) and ensure Well Specific Operating Guidelines (WSOGs) are developed for operations at every well and location. A MODU attached to the seafloor of the U.S. OCS should be operated in accordance with the appropriate WSOG.

(c) DP Operating Engineering Guidelines (Same for All Electric Ships)

Marine Technology Society Dynamic Positioning Committee (MTS-DPC) Guidelines Sept. 2012

MTS DP Part 1

4.16.2 Engineers – There should be sufficient licensed engineers on board for all expected operations.

4.16.3 At least one licensed engineer should be available at all times, should be on watch during critical activities and should have at least 6 months experience on similar equipment and operations.

4.16.4 The engineer should be fully cognizant of DP operations, familiar with the vessel's DP FMEA document and the effects of failures of equipment relating to the position keeping of the vessel.

4.16.5 In DP 2 or 3 operations, the engineer should be familiar with the general philosophy of redundancy as it relates to split mechanical, electrical and ancillary systems.

4.16.6 Electrician/Electrical Engineer – If required on board, an electrician should have appropriate high voltage training/certification, if applicable to the vessel. As with vessel engineers, the electrician/electrical engineer should have at least 6 months experience on similar equipment and operations.

4.16.7 The electrician should likewise be fully cognizant of DP operations, familiar with the vessel's DP FMEA document and the effects of failures of equipment relating to the position keeping of the vessel. Note: Where the minimum experience requirements cannot be met a risk based approach should be taken to determine the suitability of personnel and any additional support requirements for intended operations.

DNV Part 4, Chapter 8

ELECTRICAL INSTALLATIONS

DNV REQUIREMENTS:

DNV-Section-2-103 *System earthing (grounding)*

(a) System earthing (grounding) shall be effected by means independent of any earthing (grounding) arrangements of the non-current carrying parts.

(b) Any earthing (grounding) impedances shall be connected to the hull. The connection to the hull shall be so arranged that any circulating current in the earth connections do not interfere with radio, radar, communication and control equipment circuits. (IACS UR E11 2.1.4)

(c) If the system neutral is connected to earth, suitable disconnecting links or terminals shall be fitted so that the system earthing (grounding) may be disconnected for maintenance or insulation resistance measurement. Such means shall be for manual operation only.

(d) If the system neutral is connected to earth at several points, equalizing currents in the neutral earthing (grounding) exceeding 20% of the rated current of connected generators or transformers is not acceptable. Transformer neutrals and generator neutrals shall not be simultaneously earthed in the same distribution system at same voltage level. On distribution transformers with star connected primary side, the neutral point shall not be earthed.

(e) In any four wire distribution system the system neutral shall be connected to earth at all times without the use of contactors.

(f) Combined PE (protective earth) and N (system earth) is allowed between transformer/generator and N bus bar in first switchboard where the transformer secondary side/generator is terminated i.e. TN-C-S system. There shall be no connection between the N- and PE-conductor after the PEN-conductor is separated.

(g) In case of earth fault in high voltage systems with earthed neutral, the current shall not be greater than full load current of the largest generator on the switchboard or relevant switchboard section and not less than three times the minimum current required to operate any device against earth fault. Electrical equipment in directly earthed neutral or other neutral earthed systems shall withstand the current due to single phase fault against earth for the time necessary to trip the protection device. It shall be assured that at least one source neutral to ground connection is available whenever the system is in the energized mode. For divided systems, connection of the neutral to the earth shall be provided for each section. (IACS UR E11 2.1.5 and 2.1.2)

DNV-Section 6-207 *Harmonic distortion*

(a) Equipment producing transient voltage, frequency and current variations shall not cause malfunction of other equipment on board, neither by conduction, induction or radiation.

(b) In distribution systems the acceptance limits for voltage harmonic distortion shall correspond to IEC 61000-2-4 Class 2. (IEC 61000-2-4 Class 2 implies that the total voltage harmonic distortion shall not exceed 8%.) In addition, no single order harmonic shall exceed 5%.

(c) The total harmonic distortion may exceed the values given in b) under the condition that all consumers and distribution equipment subjected to the increased distortion level have been designed to withstand the actual levels. The system and components ability to withstand the actual levels shall be documented.

(d) When filters are used for limitation of harmonic distortion, special precautions shall be taken so that load shedding or tripping of consumers, or phase back of converters, do not cause transient voltages in the system in excess of the requirements in 204. The generators shall operate within their design limits also with capacitive loading. The distribution system shall operate within its design limits, also when parts of the filters are tripped, or when the configuration of the system changes.

DNV-Section 6–204 Transformer *Parallel operation*

Transformers for parallel operation shall have compatible coupling groups and voltage regulation, so that the actual current of each transformer will not differ from its proportionate share of the total load by more than 10% of its full load current.

Remark: "USCG regulation for DP system's Critical Activity Mode of Operation (CAMO) and ensure Well Specific Operating Guideline (WSOGs) are developed for operations at every well and location."

This is also the application for ships, as ships are custom-designed for specific operational requirements. Therefore the preceding requirement can be interpreted as "all vessel's critical systems Critical Mode of Operation (CAMO) must be well defined which is extension to the normal operation to ensure ship specific operating guideline." For example, electrical propulsion-related variable frequency drive operation at a critical mode of operation. This CAMO should be a part of the ship-board electrical system operational FMEA.

1.19 IEC/ISO/IEEE 80005-1-2012: UTILITY CONNECTIONS IN PORT—HIGH VOLTAGE SHORE CONNECTION (HVSC) SYSTEMS—GENERAL REQUIREMENTS

General requirements for high-voltage shore connection systems outline the connection details for RORO cargo ships, RORO passenger ships, cruise ships, container ships, LNG ships, and tankers. This standard is for the design, installation and testing of high-voltage shore connection systems for tankers and cruise ships detailing the following:

- High-voltage shore distribution system
- Shore-to-ship connection and interface equipment
- Transformer and reactors
- Semiconductor/rotating converters
- Ship distribution system
- Control, monitoring, interlocking, and power management systems

Chapter 10 provides further details of high-voltage shore power connection and low-voltage shore power connection in view of shipboard detail design and operation.

1.20 MIL STANDARD 1399 MEDIUM VOLTAGE POWER SYSTEM CHARACTERISTICS

Table 1.2   MIL-STD-1399—680: Power System Characteristics

Characteristics	5 kV Class	8.7 kV Class	15 kV Class
1. Nominal frequency	60 Hz	60 Hz	60 Hz
2. Frequency tolerances	±3%	±3%	±3%
3. Frequency modulation	$\frac{1}{2}\%$	$\frac{1}{2}\%$	$\frac{1}{2}\%$
4. Frequency transient	±4%	±4%	±4%
5. The worst-case frequency excursion from nominal frequency resulting from item 2, item 3, and item 4 combined, except under emergency conditions	±5.5%	±5.5%	±5.5%
6. Recovery time from items 4 or 5	2 sec	2 sec	2 sec
7. Nominal user voltage	4.16 kV rms	6.6 kV rms	11 kVrms, 13.8 kV rms
8. Line voltage unbalance	3%	3%	3%

Table 1.2   (*Cont.*)

Characteristics	5 kV Class	8.7 kV Class	15 kV Class
9a. Average of the three line-to-line voltages	±5%	±5%	±5%
9b. Any one line-to-line voltage, including items 8 and 9a	±7%	±7%	±7%
10. Voltage modulation	2%	2%	2%
11. The maximum departure voltage resulting from item 8, 9a, 9b, and 10 combined, except under transient or emergency conditions	±6%	±6%	±6%
12. Voltage transient tolerance	±16%	±16%	±16%
13. Worst-case voltage excursion from nominal user voltage resulting from items item 8, 9a, 9b, 10, and 12 combined, except under transient or emergency conditions	±20%	±20%	±20%
14. Recovery time from items 12 or 13	2 sec	2 sec	2 sec
15, Voltage spike	60 kV peak	75 kV peak	90 kV peak
16. Maximum total harmonic distortion	5%	5%	5%
17. Maximum single harmonic	3%	3%	3%
18. Maximum deviation factor	5%	5%	5%

Table 1.3   MIL-STD-1399—680: High-Voltage Class, and Hi-Pot Test Voltage and BIL

Class	System Voltage Maximum	System Voltage Nominal	HI-Pot Test Voltage (2 X Nominal + 1000 V)	Basic Impilse Test (BIL)
I	5.0 kV rms	4.16 kV rms	9.32 kV	30.0 kV
II	8.7 kV rms	6.6 kV rms	14.2 kV	60.0 kV
III	15.0 kV rms	11.0 kV rms	28.6 kV	90.0 kV
		13.8 kV rms		

1.21 SHIPBOARD POWER QUALITY AND HARMONICS (SEE CHAPTER 7 FOR DETAIL REQUIREMENTS)

1.21.1 IEEE Std 45-2002, Clause 4.6, Power Quality and Harmonics

Solid state devices such as motor controllers, computers, copiers, printers, and video display terminals produce harmonic currents. These harmonic currents may cause additional heating in motors, transformers, and cables. The sizing of protective devices should consider the harmonic current component. Harmonic currents in nonsensically current waveforms may also cause EMI and RFI. EMI and RFI may result in interference with sensitive electronics equipment throughout the vessel.

Isolation, both physical and electrical, should be provided between electronic systems and power systems that supply large numbers of solid state devices, or significantly sized solid state motor controllers. Active or passive filters and shielded input isolation transformers should be used to minimize interference. Special care should be given to the application of isolation transformers or filtering as the percentage of power consumed by solid state power devices compared with the system power available increases. Small units connected to large power systems exhibit less interference on the power source than do larger units connected to the same source. Solid state power devices of vastly different sizes should not share a common power circuit. Where kilowatt ratings differ by more than 5 to 1, the circuits should be isolated by a shielded distribution system transformer. Surge suppressers or filters should only be connected to power circuits on the secondary side of the equipment power input isolation transformers.

Notes:

1. To preclude radiated EMI, main power switchboards rated in excess of 1 kV and propulsion motor drives should not be installed in the same shipboard compartment as ship service switchboards or control consoles. (This is per IEEE 45-1998 Clause 4.6.)

2. To reduce the effect of radiated EMI, special considerations on filtering and shielding should be exercised when main power switchboards and propulsion motor drives are installed in the same shipboard compartment as ship service switchboards or control consoles.

3. IEEE Std 519^{TM}-1992 provides additional recommendations regarding power quality. IEEE Std 519-2014 is the latest edition, which is widely different from the 1992 version. Reference to both standards will be necessary to establish the state of the requirement and apply as recommended.

1.21.2 Power Conversion Equipment-Related Power Quality

1.21.2a IEEE Std 45-2002, Clause 31.8, Propulsion Power Conversion Equipment (Power Quality)

The following quote is an extract referring only to the power quality portion of this clause.

Whenever power converters for propulsion are applied to integrated electric plants, the drive system should be designed to maintain and operate with the power quality of the electric plant. The effects of disturbances, both to the integrated power system and to other motor drive converters, should be regarded in the design. Attention should be paid to the power quality impact of the following:

(a) Multiple drives connected to the same main power system.

(b) Commutation reactance, which, if insufficient, may result in voltage distortion adversely affecting other power consumers on the distribution system. Unsuitable matching of the relation between the power generation system's sub-transient reactance and the propulsion drive commutation impedance may result in production of harmonic values beyond the power quality limits.

(c) Harmonic distortion can cause overheating of other elements of the distribution system and improper operation of other ship service power consumers.

(d) Adverse effects of voltage and frequency variations in regenerating mode.

(e) Conducted and radiated electromagnetic interference and the introduction of high-frequency noise to adjacent sensitive circuits and control devices. Special consideration should be given for the installation, filtering, and cabling to prevent electromagnetic interference . . . .

1.22 USCG PLAN SUBMITTAL REQUIREMENTS

CFR 46-PART-10-1.15. ELECTRICAL PLAN SUBMITTAL REQUIREMENTS:

1.15.1 USCG REGULATION: (Partial Listing)

USCG CODE OF FEDERAL REGULATIONS CFR 46 PART 110

Subpart 110.25—Plan Submittal

§110.25-1 Plans and information required for new construction.

The following plans, if applicable to the particular vessel, must be submitted for Coast Guard review in accordance with §110.25-3:

(a) Elementary one-line wiring diagram of the power system, supported, by cable lists, panelboard summaries, and other information including—
 (1) Type and size of generators and prime movers;
 (2) Type and size of generator cables, bus-tie cables, feeders, and branch circuit cables;
 (3) Power, lighting, and interior communication panelboards with number of circuits and rating of energy consuming devices;
 (4) Type and capacity of storage batteries;
 (5) Rating of circuit breakers and switches, interrupting capacity of circuit breakers, and rating or setting of overcurrent devices;
 (6) Computations of short circuit currents in accordance with Subpart 111.52; and

(7) Overcurrent protective device coordination analysis for each generator distribution system of 1500 kilowatts or above that includes selectivity and shows that each overcurrent device has an interrupting capacity sufficient to interrupt the maximum asymmetrical short-circuit current available at the point of application.

(b) Electrical plant load analysis including connected loads and computed operating loads for each condition of operation.

(c) Elementary and isometric or deck wiring plans, including the location of each cable splice, a list of symbols, and the manufacturer's name and identification of each item of electrical equipment, of each—

   (1) Steering gear circuit and steering motor controller;

   (2) General emergency alarm system;

   (3) Sound-powered telephone or other fixed communication system;

   (4) Power-operated boat winch;

   (5) Fire detecting and alarm system;

   (6) Smoke detecting system;

   (7) Electric watertight door system;

   (8) Fire door holding systems;

   (9) Public address system and

   (10) Manual alarm system

(d) Deck wiring or schematic plans of power systems and lighting systems, including symbol lists, with manufacturer's name and identification of each item of electric equipment, and showing:

   (1) Locations of cables;

   (2) Cable sizes and types;

   (3) Locations of each item of electric equipment;

   (4) Locations of cable splices.

(e) Switchboard wiring diagram.

(f) Switchboard material and nameplate list.

(g) Elementary wiring diagram of metering and automatic switchgear.

(h) Description of operation of propulsion control and bus transfer switchgear.

1.23   ABS RULES FOR BUILDING AND CLASSING STEEL VESSELS (PARTIAL LISTING)

ABS-SVR - Part 4, Chapter 8, Section 1

Vessel Systems and Machinery Electrical Systems General Provisions 4-8-1

   5 Plans and Data to be Submitted

   5.1 System Plans

   5.1.1 One-Line Diagram

   5.1.2 Schematic diagrams

Schematic Diagrams Schematic diagrams for the following systems are to be submitted. Each circuit in the diagrams is to indicate type and size of cable, trip setting and rating of circuit protective device, and rated capacity of the connected load.

General lighting, normal and emergency

Navigation lights

Interior communications

General emergency alarm

Intrinsically safe systems

Emergency generator starting

Steering gear system

Fire detection and alarm system

Short-Circuit Data

Maximum calculated short-circuit current values, both symmetrical and asymmetrical values, available at the main and emergency switchboards and the down stream distribution boards.

Rated breaking and making capacities of the protective devices. Reference may be made to IEC Publication 61363-1 Electrical Installations of Ships and Mobile and Fixed Offshore Units – Part 1: Procedures for Calculating Short-Circuit Currents in Three-Phase A.C.

Protective Device Coordination Study:

This is to be an organized time-current study of all protective devices, taken in series, from the utilization equipment to the source, under various conditions of short circuit. The time-current study is to indicate settings of long-time delay tripping, short-time delay tripping, and instantaneous tripping, as applicable. Where an overcurrent relay is provided in series and adjacent to the circuit protective devices, the operating and time-current characteristics of the relay are to be considered for coordination. Typical thermal withstanding capacity curves of the generators are to be included, as appropriate.

1.24 DESIGN VERIFICATION AND VALIDATION

1.24.1 Design Verification Test Procedure (DVTP)

Table 1.4   Design Verification Test Procedure (DVTP)

	Propulsion Transformer							
Step	Equipment	Action	Result	Alt Action	Alarm	Verified	Verify
Prop transformer	Winding temp	Monitoring temp	C-degree	Meter	Hi point	Local/ remote alarm	Verified-Degrade overall operation

Table 1.4   Alternating Current (AC) LV System Power Characteristics Per Mil-STD-1399

Characteristics	Limits
Frequency	
(a) Nominal frequency	50/60 Hz
(b) Frequency tolerances	$\pm\,3\%$
(c) Frequency modulation	$\frac{1}{2}\%$
(d) Frequency transient:	
(1) Tolerance	$\pm\,4\%$
(2) Recovery time	2 sec
(e) The worst-case frequency excursion from nominal frequency resulting from item (b), item (c), and item (d) (1) combined, except under emergency conditions.	$\pm\,5\frac{1}{2}\%$
Voltage	
(a) User voltage tolerance:	
(1) Average of the three line-to-line voltages	$\pm5\%$
(2) Any one line-to-line voltage, including item (a) (1) and line voltage unbalances item (b)	$\pm7\%$
(b) Line voltage unbalance	3%
(c) Voltage modulation	5%
(d) Voltage transient:	
(1) Voltage transient tolerances	$\pm16\%$
(2) Voltage transient recovery time	2 s
(e) Voltage spike (peak value includes fundamental)	±2500 V (380–600 V) system; 1000 V (120–240 V) system
(f) The maximum departure voltage resulting from item (a) (1) and item (d) combined, except under transient or emergency conditions.	$\pm6\%$
(g) The worst case voltage excursion from nominal user voltage resulting from item (a) (1), item (a) (2), and item (d) (1) combined, except under emergency conditions.	$\pm20\%$
Waveform voltage distortion	
(a) Maximum total harmonic distortion	5%
(b) Maximum single harmonic	3%
(c) Maximum deviation factor	5%

MIL-STD—1399-680: Table-II Electrical Power System Characteristics At The Interface (High Voltage For Shipboard Application) (Partial Extract)

Table 1.5 Qualitative Failure Analysis (QFA) Sample Table

Propulsion Power—Qualitative Failure Analysis (QFA)			Propulsion Generation			
1	2	3	4	5	6	7
ITEM No	Failure description	Systems effect due to the failure	Failure detection and indication	Effect on overall ship performance due to failure	Corrective action	Failure evaluation
Propulsion Generation	Overvoltage	Actual value failure	Alarm	Probably yes	Reduce load if possible	Verify the cause of the failure. If possible correct the failure on changeover to standby generator

1.24.2 Qualitative Failure Analysis (QFA)

1.24.3 IEEE 519 Harmonic Standard

IEEE 519-1992 has been superseded by IEEE 519-2014. IEEE 519-2014 has major changes to the voltage THD requirements. Many technical details of IEEE 519-1992 have been removed. It will be necessary to refer to both versions for a better understanding of shipboard harmonic issues. For additional details refer to Chapter 7 Shipboard power system quality of power management and verification.

1.25 REMARKS FOR VFD APPLICATIONS ONBOARD SHIP

For all shipboard ungrounded power generation and distribution systems, if VFD is used, where there is a possibility of a system-level ground path, such as HRG, EMC filter, and harmonic filters where transient power-system noise is prevalent, local manual close-and-open operation of the circuit breaker with a remote manual operator and remote automatic operator is encouraged due to the fact that there may be an arcing path through that point, which may cause an explosion. Therefore, it is strongly recommended that the circuit breaker must have the capability of remote operation to avoid manual operation on a live bus. The remote operations shall be supported by mimic display. FMEA must be done for the remote circuit breaker control to establish that all required safety interlocks and functions are properly maintained. The remote circuit breaker operational capability shall be in addition to system coordination and protection features.

System-level capacitance monitoring and management is recommended.

If there are HRGs used in an ungrounded system, the total probable current path shall be monitored and managed to establish and maintain a safe level.

All high-frequency harmonics and low-frequency harmonics must be calculated and managed to establish safe harmonic levels. This recommendation may require harmonic calculation beyond the 49th harmonic of IEEE 519 requirements.

Chapter 2

Electrical System Design Fundamentals and Verifications

2.0 INTRODUCTION

Shipboard electrical power-system design fundamentals for traditional low-voltage power generation and distribution are well established and supported by rules and regulations. Marine electrical design engineers are very familiar with USCG regulations, American Bureau of Shipping (ABS) rules, and IEEE 45 recommendations, as well other national and international requirements. Commercial ship design with N + 1 redundant power generation requirements supported by emergency generation is a basic requirement for a successful shipboard electrical power system design.

Shipboard electrical design fundamentals have changed with complex operational requirements, such as power generation to support propulsion.

The equipment supply base has changed due to international market competition. The harmonization of international requirements has also created a complex mix-and-match situation, where the design and development process must be supported by verification of the design, verification of equipment selection, and verification of proper installation and shipboard operational testing.

Special requirements for offshore mobile platforms and support vessels may not be addressed by shipboard electrical design and development requirements. Offshore vessels are designed mainly according to IEC standards or a combination of US and IEC standards.

The IEC unit of measurement is metric, and the North American unit of measurement is English (NEMA). The differences between metric versus English measurement needs to be carefully addressed during the design and procurement phase. There are cases where IEC rated equipment have to be derated to meet NEMA requirements.

Shipboard Power Systems Design and Verification Fundamentals, First Edition. Mohammed M. Islam.
© 2018 the Institute of Electrical and Electronics Engineers, Inc. Published 2018 by John Wiley & Sons, Inc.

The THD related point of common coupling (PCC) for IEC equipment may be different from NEMA equipment.

The shipboard switchboard arcing fault has created requirements for an arc-resistant switchboard. For the non-arc-resistant switchboard, an arc detection and management system, including instantaneous current sensing, arcing light sensing detection, an alarm, and protective device operation, has become a point of design consideration.

The use of IEC switchgear with IEEE-45-rated cable, must be well understood so that the challenging point of common coupling between IEC equipment and NEMA equipment is well defined and appropriate measures are taken to make the installation safe. For example, an IEC-rated generator is provided with studs or a simple plate with holes to connect power cables. The studs are provided with specific threads in millimeters, which must match the NEMA (English) measurement of the feeder cable lug holes for IEEE 1580 cables. The stud should be supplied with nuts and washers of the appropriate size and type. Otherwise, the procurement of those accessories becomes very challenging. Failure to properly terminate the cable will create a major arcing problem. Therefore, verification is mandatory for the acceptance of final cable termination. The IEC cable termination usually has single-hole lugs. However, the use of two-hole lugs is recommended by IEEE 45 for shipboard installation to ensure integrity of the connection point. If the shipboard cable termination is not done properly, the termination will fail due to shipboard roll-pitch and vibration. IEEE Std 45-2002 Clause 7.4.4 recommends two-hole lugs for both NEMA as well as IEC cable terminations. The designer must be familiar with these as design fundamentals, and then develop procurement specifications and verify that equipment is supplied as intended. This is the only way to verify a satisfactory design process.

IEEE 45-2002 Clause 7.4.4 Terminal arrangements and incoming cables
Generators should be provided with silver or tin-plated copper fixed terminals used for connection of incoming cable and lugs or provided with copper terminal leads suitably secured to the generator frame. Terminals and terminal leads shall be adequately sized for the rating of the generator. Fixed terminals shall be supplied with suitable provisions for NEMA or IEC lugs with a minimum of two holes. Terminal lugs shall be suitable for the conductor size and temperature rating of the incoming cables. It may be necessary to provide a transition bus to accommodate metric and ANSI/NEMA dimensions.
Terminal boxes should be of sufficient size to accommodate the generator leads and terminals without crowding or exceeding the bend radius of the leads. Where shielded cables are used on generation voltages of 3 kV and higher, sufficient straight space shall be provided between the point of cable entrance and terminals for installation of stress cones. Space should also be provided for current transformers used for differential, other protection, or metering. Additional space may be necessary for bus bars and other means of interconnection of neutral leads for six lead machines. Each box should be of adequate mechanical strength and rigidity to protect the contents and to prevent distortion under all normal conditions of service.

If the equipment is already supplied, field modification may not be permitted. Therefore, there will be a need for an alternate acceptance process.

IEEE 45-2002 Clause 4.4 Selection of voltage and system type
For small vessels having minimal power apparatus (up to 15 kW), 120 V, three-phase
or single-phase generators may be used, with 11 5V, three-phase or single-phase
distribution for power and lighting. Single- phase lighting feeders should be balanced at
the switchboard to provide approximately equal load on the three-phase system.
For intermediate-size vessels with power apparatus (up to 100 kW), generators may
be 230 V, three-phase or 240 V, three-phase; the power utilization at 220 V or 230 V
three-phase, respectively; and lighting distribution at 120 V, three-phase, three-wire or
208/120 V three-phase, four-wire. Power and lighting utilization should be at 200/115 V,
three-phase.
For large vessels of a size and type that require a dual-voltage system (two systems
isolated by transformers operating at different voltages), first consideration should be
given to 450 V, 480 V, 600 V, or 690 V generation with power utilization at 44 0V, 460 V,
575 V, or 660 V, respectively, and lighting distribution at 120 Vor 230 V three-phase,
three-wire or 120/208 V, three-phase, four-wire.
For vessels having a very large electrical system requiring higher voltage power
generation, consideration should be given to generating at 13,800 V, 11,000 V, 6600 V,
4160 V, 3300 V, or 2400 V with some power utilization at 13,200 V, 10,600 V, 6000 V,
4000 V, 3150 V, or 2300 V, three-phase, respectively, with lower utilization voltages to
be derived from transformers.
For vessels requiring DC power generation, and having little power apparatus, 120 V
DC generators are recommended with a 115 V DC lighting and power distribution
system. Where an appreciable amount of DC power apparatus is provided, 240/120 V
DC, three-wire generators and 230 V DC power distribution system and 230/115 V DC
three-conductor lighting feeders may be selected. Branch circuits from lighting panel
boards should be 115 V DC two-wire.

This is to show and analyze various shipboard electrical designs with electrical propulsion with variable frequency drive. These notes relate directly to various component requirements and some rational to verify specific applications. This is developed for the design engineers so that the design is in compliance with specific guidelines.

2.1 DESIGN BASICS

Shipboard electrical system design and development must follow an established process as follows:

- Must understand the requirements
- Must develop multiple design options so that those options are properly documented and evaluated and selected
- The acceptable option of the selection process must be documented
- The design process must be validated to ensure design optimization
- All design phases must be properly defined. Any undefined design issue will call for a corrective undertaking and so may lead to prohibitive cost or to timing

that may not be appropriate to adapt any change. Also, there may not be any space available to add new equipment.

2.2 MARINE ENVIRONMENTAL CONDITION REQUIREMENTS FOR THE SHIPBOARD ELECTRICAL SYSTEM DESIGN

The shipboard ungrounded electrical power system, as the ship floats in the water, requires special environmental considerations. These considerations are well defined in IEEE Std 45-2002 Clause 1.5, as given here:

IEEE Std 45-2002 Clause 1.5.1 Normal design and operating conditions

Systems and equipment should be suitable for continuous operation under the following shipboard conditions:

(a) Exposure to moisture-laden and salt-laden atmosphere, weather, sun, high wind velocities, and ice.

(b) Equipment and systems shall be designed for temperature extremes and conditions expected. Typically, the following conditions can be used: ambient temperature values of 40 °C in accommodation areas, and similar spaces; 45 °C in main and auxiliary machinery spaces; 50 °C for rotating machinery and propulsion equipment in main and auxiliary machinery spaces containing significant heat sources such as prime movers and boilers; and 65 °C in the uptakes of machinery spaces containing prime movers and boilers; all at relative humidity up to 95%. The design value for seawater cooling temperature should be 32 °C.

(c) Roll and pitch of a vessel underway, as shown in Table 1.

(d) Vibration of a vessel underway: Electrical equipment should be constructed to withstand at least the following:

  1. Vibration frequency range of 5 to 50 Hz with a velocity amplitude of 20 mm/s.

  2. Peak accelerations due to ship motion in a seaway of ± 5.9 m/sec^2 for ships exceeding 90 m in length, and ± 9.8 m/sec^2 for smaller ships, with a duration of up to 10 s.

Table 2.1   IEEE Std 45-2002 Clause 1.5 Table 1 For Roll and Pitch Requirements

	Roll		Pitch	
	Static (°)	Dynamic (°)	Static (°)	Dynamic (°)
Ship service equipment	15	22.5	5	75
Emergency equipment*	22.5	22.5	10	10
Switchgear	45	45	45	45

[a] vessels designed for carriage of liquefied gases and of chemicals, the emergency power installation is to remain operable with the vessel flooded to its permissible athwartship inclination up to a maximum of 30°.

IEEE Std 45 Clause 1.5.2 Abnormal design and operating conditions

Special conditions for a specific ship design or operating arrangement may require special consideration. Examples of such conditions include, but are not necessarily limited to

- Exposure to damaging fumes or vapors, excessive or abrasive dust, steam, salt-spray, ice, sunlight, physical damage, and so on
- Exposure to high levels of shock and vibration
- Exposure to high or low temperatures
- Operation in flammable atmospheres
- Exposure to unusual loading or unloading conditions affecting list and trim
- Unusual operating cycles, frequency of operation, poor power quality, special insulation requirements, stringent or difficult maintenance requirements, and so on

2.3 POWER SYSTEM CHARACTERISTICS: MIL-STD-1399 POWER REQUIREMENTS

1.1 Scope. This military standard section establishes electrical interface characteristics for shipboard equipment utilizing AC electric power to ensure compatibility between user equipment and the electric power system. Characteristics of the electric power system are defined and tolerances are established, as well as requirements and test methods for ensuring compatibility of shipboard user equipment with the power system. The policies and procedures established by MIL-STD-1399 are mandatory. This section and the basic standard are to be viewed as an integral single document for use in the design and testing of electric power systems and user equipment.

1.2 Classification. Types of shipboard electric power to be supplied from the electric power system are classified as low voltage as follows:

Type I – Type I power is 440 or 115 volts (V), 60 hertz (Hz) ungrounded and is the standard shipboard electric power source. Type I power is used unless a deviation is granted.

Type II – Type II power is 440 or 115 V, 400 Hz ungrounded and has only limited application. Use of Type II power requires the submittal and approval of a deviation request.

Type III – Type III power is 440 or 115 V, 400 Hz ungrounded having tighter tolerances as compared to Type II. Type III power has restricted use and its use requires the submittal and approval of a deviation request.

1.2.1 Special power classification for avionic shops and aircraft servicing. Types of shipboard electric power supplied only for avionic shops and aircraft servicing are as follows:

Type I – Type I power is 115/200 V, 60 Hz, 3-phase, 4-wire, wye-grounded. This power is only provided for avionic shops. Type III – Type III power is 115/200 V, 400 Hz, 3-phase, 4-wire, wye-grounded. This power is only provided for avionic shops and for aircraft servicing.

1.2.3 Special power classification for NATO load equipment. Types of shipboard electric power supplied only for NATO load equipment are as follows:

Type I – Type I power is 230 V, 60 Hz, 3-phase, ungrounded or 230 V, 60 Hz, single-phase, grounded or ungrounded. Its tolerances are the same as for Type I power as described in Table I except that the spike voltage will be at 1400 V peak.

1.2.3 Special non-standard power. For types of shipboard electric power supplied for specific industrial equipment such as washers, dryers, etc., see NAVSEA Drawings 7512881 for 120/208 Vrms loads and 7598285 for 120/240 Vrms loads. Non-standard power should comply with Type I tolerances.

NOTE for Human Body Leakage current voltage and frequency limits.

1.3 Electrical interface. The basic characteristic and constraint categories concerned with this interface are shown symbolically on Figure 1. This interface is a location between the electric power system and the user equipment. The interface is at the junction where the cable designations change from power or lighting designations, such as P, EP, PP, L, EL, or SF, to other designations or where no cable designation changes are made at the user equipment electric power input terminals. Functionally, the interface is the location wherein the electric power system characteristics (see 5.1) and the user equipment constraints apply.

Section-5.1.1.1 Type I, 60 Hz power. The ship service electrical power distribution system supplied by the ship's generators is 440 Vrms, 60 Hz, three-phase, ungrounded. Power for the ship's lighting distribution system and other user equipment such as electronic equipment, supplied from the ship service power distribution system through transformers, is 115 Vrms, 60 Hz, three-phase, ungrounded. Single-phase power is available from both the 440 Vrms and the 115 Vrms systems. The ship service power and lighting distribution systems are labeled as Type I. Type I, 230 Vrms, 60 Hz, single or three-phase, grounded or ungrounded power can be made available for NATO load equipment upon special request.

2.4 ABS TYPE APPROVAL PROCEDURE (TAKEN FROM ABS DIRECTIVES)

ABS type approval directive is for ABS approved products for marine application. This program is a process of evaluation, audit and maintenance of ABS approved product database.

A Type Approved Product has satisfied the processes of:

1. An engineer's evaluation of a design to determine conformance with specifications. The manufacturer should submit sufficient information to allow ABS to

Table 2.2   Extract of Type I Power Requirements from MIL-STD-1399 (Partial) – Low Voltage

Characteristics	Type I
FREQUENCY	
a. Normal Frequency	60 Hz
b. Frequency Tolerance	Plus or minus 3%
c. Frequency Modulation	$\frac{1}{2}$%
d. Frequency Transient	
1. Tolerance	Plus or minus 4%
2. Recovery time	2 seconds
e. Worst case, frequency excursion from nominal frequency Resulting from b, c, and d(1) combined except under Emergency condition	Plus or minus 5-1/2%
VOLTAGE	
f. Normal user voltage	440, 115 V
g. User voltage tolerance	
1. Average of three line-to-line voltages	Plus or minus 5%
2. Any one line-to-line voltage including g(1) and Line voltage unbalance (h)	Plus or minus 7%
h. Line voltage unbalance	3%
i. Voltage modulation	2%
j. Voltage transient	
(1) Voltage transient tolerance	Plus or minus 16%
(2) Voltage transient recovery time	2 seconds
k. Voltage spikes (peak value, includes fundamental)	Plus minus 2.500 V for 440 V sys and 1000 V for 115 V
l. Maximum departure voltage resulting from g(1), g(2), (h), and (I) combined, except under transient or emergency conditions	
m. The worst case voltage excursion from nominal user voltage resulting from (g)(1), (g)(2), (h), (I), and combined except under emergency conditions.	
n. Insulation resistance test	
(a) Surface ships	500 VDC
WAVEFORM (VOLTAGE)	
(o) Maximum total harmonic distortion	5%
(p) Maximum single harmonic	3%
(q) Maximum deviation factor	5%
EMERGENCY CONDITIONS	
(r) Frequency excursion	Minus 100 to plus 12%
(s) Duration of frequency excursion	Up to 2 minutes
(t) Voltage excursion	Minus 100 to plus 35%
(u) Duration of voltage excursion	
(1) lower limit (minus 100%)	Up to 2 minutes
(2) upper limit (plus 35%)	2 minutes

Table 2.3   Extract of Type I Power Requirements from MIL-STD-1399 (Partial)—Medium Voltage

Characteristics	5 kV Class	8.7 kV Class	15 kV Class
Frequency			
1. Nominal Frequency	60 Hz	60 Hz	60 Hz
2. Frequency Tolerance	3%	3%	3%
3. Frequency Modulation	0.50%	0.50%	0.50%
4. Frequency Transient Tolerance	±4%	±4%	±4%
5. Worst case frequency excursion from nominal frequency resulting from items 2, 3, and 4 combined, except under emergency conditions	±5.5%	±5.5%	±5.5%
6. Recovery time from 4 or 5	2 sec	2 sec	2 sec
Voltage			
7. Nominal user voltage	4.16 kV rms	6.6 kV rms	11 kV rms
8. Line voltage unbalance	3%	3%	3%
9. User voltage tolerance			
9a. Average of three line-to-line voltages	±5%	±5%	±5%
9b. Any one line-to-line voltage, including items 8 and 9a			
10. Voltage modulation	2%	2%	2%
11. Maximum departure voltage resulting from items 8, 9a, 9b, and 10 combined, except under transient or emergency conditions	±6%	±6%	±6%
12. Voltage transient tolerance	±16%	±16%	±16%
13. Worst case voltage excursion from nominal user voltage resulting from items 8, 9a, 9b, and 10 combined except emergency conditions	±20%	±20%	±20%
14. Recovery time from items 12 or 13	2 Sec	2 Sec	2 Sec
15. Voltage spike	60 kV peak	75 kV peak	95 kV peak
Wave form (Voltage)			
16. Maximum total harmonic distortion	5%	5%	5%
17. Maximum single harmonic	3%	3%	3%
18. Maximum deviation factor	5%	5%	5%
Emergency Conditions			
19. Frequency excursion	(−100 to 12%)	(−100 to 12%)	(−100 to 12%)
20. Duration of frequency excursion	2 min	2 min	2 min
21. Voltage excursion	(−100 to 35%)	(−100 to 35%)	(−100 to 35%)
22. Duration of voltage excursion			
22a. Upper limit (+35%)	2 min	2 min	2 min
22b. Lower limit (−100%)	2 min	2 min	2 min

determine if the product meets specification. This results in a Product Design Assessment Certificate (PDA).

2. Witnessing manufacture and testing of a type of the product to determine compliance with the specification.

3. A surveyor's evaluation of the manufacturing arrangements to confirm that the product can be consistently produced in accordance with the specification. This results in the issue of a Manufacturing Assessment Certificate.

Basically a PDA is the pre-approval of a product for use on a variety of ABS class ships. The PDA reduces the turn around time for approval on a specific ship. When a specific ship is chosen, the technical staff would then verify that the product, as already approved, is suitable for the intended use. This can be done with a simple review of the PDA and not require submittal of further documentation from the manufacturer. In the situation of a product that does not require Unit Certification; the manufacturer may be served by having a PDA only. Note that this PDA normally would have a five-year validity, subject to Rule changes, design changes, etc.

Any manufacturer may apply for a Type Approved Product. However, a Type Approved Product is of more benefit to the manufacturer when there is a requirement for Unit Certification and an ABS surveyor attends for material testing and other tests during manufacturer. Being a Type Approved Product, expedites the Unit Certification process.

2.4.1 List of Recognized Laboratories

The following list and links to internet sites are provided as a courtesy only. ABS is not a certifying laboratory and cautions that users of a laboratory double check the current certification of any chosen laboratory. ABS cannot be responsible for the accuracy of these lists and retains the right to review all information presented from a laboratory prior to accepting the data as pedigree of the product presented for ABS evaluation.

- IMO List of Laboratories
- US Department of Labor, Nationally Recognized Testing Laboratories
- USCG Listing of Labs Approved to Conduct Testing for the US Administration
- United Kingdom Accreditation Services (UKAS) Accredited Laboratories
- International Accreditation Service, Inc.

2.4.2 Nationally Recognized Testing Laboratory Program

Welcome to the Nationally Recognized Testing Laboratory (NRTL) Program. Workplace product safety is a critical component of workplace safety and both the construction and general industry OSHA electrical standards contain requirements for certain products to be tested and certified by an NRTL. NRTLs are private sector organizations that are recognized by OSHA to perform this certification. Each NRTL has a scope of test standards that they are recognized for, and each NRTL uses its own unique registered certification mark(s) to designate product conformance to the applicable product safety test standards. After certifying a product, the NRTL authorizes the manufacturer to apply a registered certification mark to the product. If the certification is done under the NRTL program, this mark signifies that the NRTL tested and certified the product, and that the product complies with the requirements of one or more appropriate product safety test standards. Users of the product can generally rely on the mark as evidence that the product complies with the applicable OSHA approval requirement(s) and is safe for use in the workplace.

Program Resources

- NRTL Program Regulations
- NRTL Program Policies, Procedures, and Guidelines (NRTL Directive - CPL 01-00-003 - CPL 1-0.3) (PDF)
- Current List of Appropriate Test Standards Under the NRTL Program
- Type of Products Requiring NRTL Approval
- Specific References to OSHA Standards Requiring NRTL Approval
- Typical Registered Certification Marks

ORGANIZATIONS CURRENTLY RECOGNIZED BY OSHA AS NRTLS

The pages below include information about the NRTL (such as the list of standards, sites, and programs that OSHA has recognized for the NRTL).

- Canadian Standards Association (CSA)
- Curtis-Straus LLC (CSL)
- FM Approvals LLC (FM)
- Intertek Testing Services NA, Inc. (ITSNA)
- MET Laboratories, Inc. (MET)
- Nemko-CCL (CCL)
- NSF International (NSF)
- QPS Evaluation Services Inc. (QPS)
- SGS North America, Inc. (SGS)
- Southwest Research Institute (SWRI)
- TUV Rheinland of North America, Inc. (TUV)
- TUV Rheinland PTL, LLC (TUVPTL)
- TUV SUD America, Inc. (TUVAM)
- TUV SUD Product Services GmbH (TUVPSG)
- Underwriters Laboratories Inc. (UL)

2.4.3 Procedure for Becoming Type Approved

Step 1: Application

IA. DRAWING AND DOCUMENT SUBMITTAL PACKAGE INCLUDES:

- Completed application
- Application (MS Word format)
- One electronic copy of applicable drawings, datasheets, test results, etc.
- Sufficient data to verify compliance with stated standards
- Must be specific to the intended services stated on the application

IB. STANDARDS FOR APPROVAL:

- ABS Rules and Guides
- US National and Foreign National
- US Federal
- Foreign Federal (Administrative)
- International (IMO) Safety (SOLAS)
- Manufacturer's standards

Step 2: Evaluation and Assessment

2A. DESIGN EVALUATION:

- ABS technical office reviews plans; where satisfied, plans are approved and a PDA (certificate) is issued
- Prototype Test

2B. MANAGEMENT ASSESSMENT:

- Performed by local ABS surveyor
- Standardized checklists for surveyor guidance
- Requirements not applicable for all clients
- Emphasis on quality control
- Based on ISO 9000 standard

2C. PRODUCTION ASSESSMENT:

- Performed by local ABS surveyor
- Standardized checklists for surveyor guidance
- Production surveillance will be conducted on an annual basis

Step 3: Type Approval Certification

In some cases a Unit Certification test by a surveyor may be required for each production unit of equipment.

3A. ISSUANCE OF CERTIFICATE:

- ABS port office issues a five-year Manufacturing Assessment Certificate to the manufacturer
- Product is included in ABS List of Type Approved Equipment and ABS website
- Manufacturer may market product as ABS Type Approved on the equipment
- Certification of Type Approval may then be printed from the website

3B. MAINTENANCE OF ABS TYPE APPROVAL:

- Annual plant surveillance
- Products may be added to listing without interim plant surveys

2.5 SHIPBOARD ELECTRICAL POWER SYSTEM DESIGN BASICS

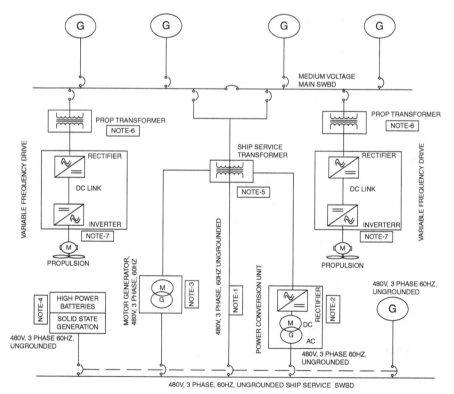

Figure 2.1 Typical Electrical One-Line Diagram with Medium-Voltage Distribution with Electric Propulsion and Type-I Distribution.

2.5.1 Table 2.4: Explanation for Note 1 of Figure 2.1 (Use of Multiple Options, Step Down Transformer, MG Set, PCU)

Table 2.4 Explanation for Note 1 of Figure 2.1 (Multiple Options)

Item	Description	Verification
1	The arrangement with step down transformer is the simplest and most economical design directly supplying power from the MV switchboard to the ship service switchboard when higher voltage generation and distribution is available onboard. This is a traditional design, in compliance with rules and regulations. However, in view of the VFD propulsion drive application the simple step down transformer may not comply with the transient stability requirements as well as the total harmonic distortion (THD) limitation of the system. Therefore in the harmonic environment, other design features must be considered and added to comply with those requirements. Some of those features are presented as options to be considered. Those options are presented in Tables 2.2 through 2.7.	The ship service distribution requirements are to provide class-1 power for certain equipment. The transformer is fed by the VFD environment which will have harmonics with various orders. The harmonics may have high-frequency characteristics. It is mandatory to supply the required type of power.
2	The ship service power generation requirement is to provide dedicated ship service generators with redundant capabilities with paralleling features as necessary. As an alternate to the ship service generators usually an MV transformer is used so that if the propulsion switchboard is energized, ship service power can be made available through the transformers. This arrangement is recommended so that environmental protection agency (EPA) requirements can be met for low emissions for specific regions.	If power requirements cannot be met with the ship service generator, then the next step will be to consider using a 480 V, 3-phase output transformer. This application will require analytical verification of the main bus harmonic level. The regulatory body requirements for the harmonic level are given in IEEE-45, typically 5% or 5%–8% for specific applications. If the verification results show any number above the acceptable THD then alternate arrangements must be made. Sometimes design progresses without such verification at the detail design stage result in a design not acceptable to the regulatory body. Then, some type of filter is used to bring the harmonics to an acceptable level.

Table 2.4   (*Cont.*)

Item	Description	Verification
3	This arrangement provides necessary galvanic isolation. Therefore all necessary requirements must be met such as: a dedicated ground detection system to support transformer secondary distribution.	
4	REMARKS:	

- The design must be verified by an authority having design jurisdiction before the design is submitted for approval.
- Usually system-level harmonic calculations should be made to establish THD in the system.
- If the THD level cannot be met with this design, then an alternate design must be considered.

2.5.2 Table 2.5: Explanation for Note 2 of Figure 2.1 (Use of Power Conversion Unit to Supply Power from MV SWBD to the Ship Service SWBD)

Table 2.5   Explanation for Note 1 of Figure 2.1 (Use of PCU-Note-2)

Item	Description	Verification
1	This is a complex design with rectifier, DC motor, and AC generator. This design provides ship service power within the boundary of clean power requirement.	This design must be verified by an authority having design jurisdiction before the design can be submitted to the approval authority. Usually system-level harmonic calculations should be made to establish THD in the system
2	Transformer-fed power is a substitute for the ship service generator. Nowadays, there are environmental protection agency (EPA) requirements for low emissions for specific regions.	If the power requirements cannot be met with a ship service generator, then the next step will be to consider using a 480 V, 3-phase output transformer. This application will require analytical verification of the main bus harmonic level. The regulatory of acceptable harmonic level typically 5% to 8% varying application to application. If the verification results show any number above the acceptable THD then an alternate arrangement must be made. Sometime design progresses without such verification at the detail design stage. If the resultant design is not acceptable to the regulatory body, then some type of filter is used to bring the harmonics to an acceptable level.

2.5.3 Table 2.6: Explanation for Note 3 of Figure 2.1 (Use of Motor Generator with MV Input to AC Motor and Driving AC Generator)

Table 2.6 Explanation for Note 1 of Figure 2.1 (Use of MG Set: Note-3)

Item	Description	Verification
1	This design is acceptable if the main switchboard can maintain the voltage transient as well as the THD to an acceptable limit. Otherwise an alternate design must be considered, such as the design explained in Note 2.	The ship service distribution requirements are to provide class-1 power for certain equipment. The transformer is fed by a VFD environment, which will have harmonics with various orders. The harmonics may have high-frequency characteristics. It is mandatory to supply the required type of power.
2		If the power requirements cannot be met with the ship service generator, then the next step will be to consider using a 480 V, 3-phase output transformer. This application will require analytical verification of the main bus harmonic level. The regulatory body requirements for the harmonic level are given in IEEE-45, typically 5% or 5%–8% for specific applications. If the verification results show any number more than the acceptable THD, then alternate arrangements must be made. Sometimes design progresses without such verification at the detail design stage and the resultant design is not acceptable to the regulatory body. then some type of filter is used to bring the harmonic to an acceptable level.

2.5.4 Table 2.7: Explanation for Note 4 of Figure 2.1 (High-Power Battery Supplying Power to the 480 V Ship Service Switchboard)

Table 2.7 Explanation for Note 1 of Figure 2.1 (Use of Battery-UPS-Note-4)

Item	Description	Verification
1	This is a newer design feature that has been gaining popularity due to tremendous progress in battery technology. The battery system with high power is available at a reasonable cost and is somewhat maintenance-free, making it suitable for shipboard application. This can easily support the ship service load for a defined duration. This design consists of a rechargeable battery bank feeding a DC bus. An inverter system converts DC power to three-phase 60 Hz for the ship service load.	The ship service distribution requirements are to provide class-1 power for certain equipment. The transformer is fed by a VFD environment, which will have harmonics with various orders. The harmonics may have high-frequency characteristics. It is mandatory to supply the required type of power.

Table 2.7   (*Cont.*)

Item	Description	Verification
2	Transformer-fed power is a substitute for the ship service generator. Nowadays, there are environmental protection agency (EPA) requirements for low emissions for specific regions.	If the power requirements cannot be met with the ship service generator, then the next step will be to consider using a 480 V, 3-phase output transformer. This application will require analytical verification of the main bus harmonic level. The regulatory body requirements for the harmonic level are given in IEEE-45, typically 5% or 5%–8% for specific applications. If the verification results show any number more than the acceptable THD, then alternate arrangements must be made. Sometime a design progresses without such verification at the detail design stage and the resultant design is not acceptable to the regulatory body. Then some type of filter is used to bring the harmonics to an acceptable level.

2.5.5 Table 2.8: Explanation for Note 5 of Figure 2.1 (Use of Step Down Service Transformer to Supply Power from MV SWBD to the Ship Service SWBD)

Table 2.8   Explanation for Note 1 of Figure 2.1 (Use of Step Down Transformers-Note-5)

Item	Description	Verification
1	This is the simplest and economical design with 3-phase in and 3-phase out. The transformer output is for ship service distribution.	The ship service distribution requirements are to provide class-1 power for certain equipment. The transformer is fed by a VFD environment, which will have harmonics with various orders. The harmonics may have high frequency characteristics. It is mandatory to supply the required type of power.
2	Transformer-fed power is a substitute for the ship service generator. Nowadays, there are environmental protection agency (EPA) requirements for low emissions for specific regions.	If the power requirements cannot be met with the ship service generator, then the next step will be to consider using a 480 V, 3-phase output transformer. This application will require analytical verification of the main bus harmonic level. The regulatory body requirements for the harmonic level are given in IEEE-45, typically 5% or 5-8% for specific applications. If the verification results show any number more than the acceptable THD, then alternate arrangements must be made. Sometime a design progresses without such verification at the detail design stage and the resultant design is not acceptable to the regulatory body. Then some type of filter is used to bring the harmonics to an

2.5.6 Table 2.9: Explanation for Note 6 of Figure 2.1 (Variable Frequency of Adjustable Drive for Electrical Propulsion Application)

Table 2.9 Explanation for Note 1 of Figure 2.1 (Use Of ASD Propulsion-Note-6)

Item	Description	Verification
1	This is the simplest and most economical design with 3-phase in and 3-phase out. The transformer output is for ship service distribution.	The ship service distribution requirements are to provide class-1 power for certain equipment. The transformer is fed by a VFD environment, which will have harmonics with various orders. The harmonics may have high-frequency characteristics. It is mandatory to supply the required type of power.
2	Transformer-fed power is a substitute for the ship service generator. Nowadays, there are environmental protection agency (EPA) requirements for low emissions for specific regions.	If the power requirements cannot be met with ship service generator, then the next step will be to consider using a 480 V, 3-phase output transformer. This application will require analytical verification of the main bus harmonic level. The regulatory body requirements for the harmonic level are given in IEEE-45, typically 5% or %–8% for specific applications. If the verification results show any number more than the acceptable THD, then alternate arrangements must be made. Sometimes a design progresses without such verification at the detail design stage and the resultant design is not acceptable to the regulatory body. Then some type of filter is used to bring the harmonics to an acceptable level.

Shipboard electrical design fundamentals have gone through major changes. Those changes must be well understood so that the operational requirements are met within the design parameters without taking any shortcuts contributing to safety compromise. Therefore the design fundamentals include a verification process at the concept stage of the design.

2.6 SHIPBOARD ELECTRICAL STANDARD VOLTAGES

IEEE 45 standard LOW VOLTAGE AC in use: 120 V, 230 V, 480 V. (All voltage-levels below 600 VAC)

IEC low-voltage level below 1000 VAC. (IEC uses 690 VAC system for low-voltage drive application)

The following voltage levels and frequency shall be used:

480/208 V/120 system shall be used as distribution voltage (3-phase and single-phase, grounded and ungrounded). 480/208 V system shall be used as distribution voltage (3-phase and single-phase, grounded and ungrounded). 480/120 system shall be used as distribution voltage (3-phase and single-phase, grounded and ungrounded). 208 V system shall be used as distribution voltage (3-phase and single-phase, grounded and ungrounded). 120 V system shall be used as distribution voltage (3-phase and single-phase, grounded and ungrounded).

480 V, 3-phase generation and distribution voltage. Should be used when total installed generator capacity is below 4 MW. Should be used for consumers below 400 kW and as the primary voltage for VFD motors.

690 V, 3-phase generation and distribution voltage. Should be used when total installed generator capacity is below 4 MW. Should be used for consumers below 400 kW and as the primary voltage for converters for VFD motors.

6.6 kV, 3-phase generation and distribution voltage. Should be used when total installed generator capacity is between 4–20 MW. Should be used for motors from 300 kW and above.

11 kV, 3-phase generation and distribution voltage. Should be used when total installed generator capacity exceeds 20 MW. Should be used for motors from 400 kW and above.

750 VDC used for operational-related motors.

480 VDC/480 VAC uninterruptable power supply used for non-break power supply to vital and emergency loads.

48 VD, 24 VDC, and 12 VDC.

There are other voltages and frequencies in use, which have been omitted for simplification.

2.6.1 NORSOK Standard 6.1 System Voltage and Frequency

Table 2.10   Shipboard Standard Voltage

11 kV, 3-Phase, 60 Hz:	Generation and distribution voltage. 11 kV generation voltage should be used when total installed power requirement exceeds 20 MW. Should be used for motors from 400 kW and above.
6.6 kV, 3-Phase, 60 Hz:	Generation and distribution voltage. 6.6 kV generation voltage should be used when total installed power requirement is between 4-20 MW. Should be used for motors from 300 kW and above.
690 V, 3-Phase, 60 Hz	Generation and distribution voltage. 690 V generation voltage should be used when total installed power requirement is below 4 MW. Should be used for consumers below 400 kW and as primary voltage for converters for drilling motors.
UPS 230 VDC & 750 VDC UPS 48 VDC	IT system shall be used as distribution voltages for instrumentation, single-phase: control, telecommunication and safety systems.

2.7 VOLTAGE AND FREQUENCY RANGE (MIL-STD-1399)

Shipboard electrical power demand has been increasing due to the transition from traditional mechanical propulsion drives to electrical propulsion drives, and auxiliary systems, such as mechanical gears and hydraulic gears, changing over to variable drive electrical systems. To meet this electrical power demand, power is being generated at higher voltages such as 480 V, 690 V, 6.6 kV, etc.

These new requirement of high voltage and high demands sophisticated system protection and sophisticated control and monitoring system. It is very important to understand the criteria for selecting generation voltage as well as distribution voltage. To select the most suitable voltage level, it is necessary to compare the various alternative solutions, considering the basic requirements (e.g., fault duty, rating current, and maximum allowable voltage drop in normal operating conditions and during large motor start-up) and taking into account the standardized equipment available on the market.

2.8 UNGROUNDED SYSTEM CONCEPT (ANSI AND IEC)

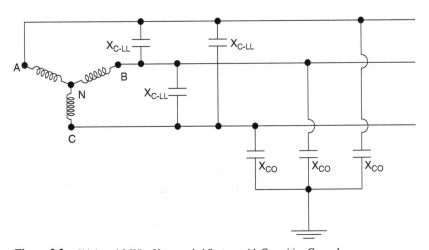

Figure 2.2 Shipboard 3-Wire Ungrounded System with Capacitive Grounds.

2.9 CONCEPT DESIGN

2.9.1 Power Generation

The customer should specify the available voltage. Normally, this will be in the range 380 V to 690 V. Higher voltage supplies offer a small advantage in that they enable smaller motors to be utilized and reduce losses due to higher currents. Indeed, there are some installations on vessels which utilize voltages of 6 to 11 kV. However, these tend to be on very heavy lift, specialized vessels which justify the production of these high voltage (and high cost) motors whereas motors in the 440 V range are readily available and utilized by hydraulic equipment manufacturers. The frequency should also be stated.

2.9.2 Power Distribution

2.10 DESIGN FEATURES OUTLINED IN

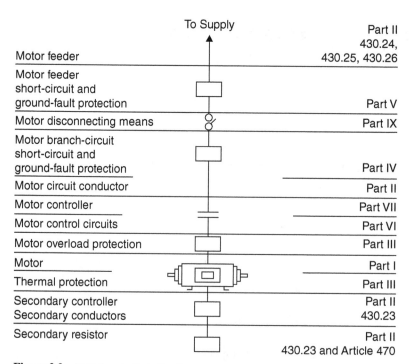

	To Supply	Part II
Motor feeder		430.24, 430.25, 430.26
Motor feeder short-circuit and ground-fault protection		Part V
Motor disconnecting means		Part IX
Motor branch-circuit short-circuit and ground-fault protection		Part IV
Motor circuit conductor		Part II
Motor controller		Part VII
Motor control circuits		Part VI
Motor overload protection		Part III
Motor		Part I
Thermal protection		Part III
Secondary controller		Part II
Secondary conductors		430.23
Secondary resistor		Part II 430.23 and Article 470

Figure 2.3   NEC Sample Protection Features.

2.11 PROTECTIVE DEVICE–CIRCUIT BREAKER CHARACTERISTICS

Code Letter	Kilovolt-Amperes per Horsepower with Locked Rotor
A	0–3.14
B	3.15–3.54
C	3.55–3.99
D	4.0–4.49
E	4.5–4.99
F	5.0–5.59
G	5.6–6.29
H	6.3–7.09
J	7.1–7.99
K	8.0–8.99
L	9.0–9.99
M	10.0–11.19
N	11.2–12.49
P	12.5–13.99
R	14.0–15.99
S	16.0–17.99
T	18.0–19.99
U	20.0–22.39
V	22.4 and up

Figure 2.4   NEC Locked Rotor Indicating Code Letters.

2.12 FAULT CURRENT CALCULATION AND ANALYSIS REQUIREMENT

The fault current analysis is performed to establish the mechanical strength and electrical properties of circuit breakers, switches, bus bars, and other equipment likely to experience the energy released by an electrical fault. This equipment must be designed either to withstand the fault energy, or in the case of circuit breakers and fuses, to open and clear the fault current promptly and safely.

The low-voltage shipboard power system utilizes air circuit breakers and molded case circuit breakers for overload as well as short-circuit protection of the electric plant. These devices are designed and set to detect unusual overcurrent in an electrical system and open the circuit if an overcurrent occurs. Electrical fault is one of the conditions that can cause an instantaneous extremely high current and analysis of the fault conditions is required to be well thought-out when designing or modifying a shipboard electrical power system.

Table 2.11   Circuit Breaker Characteristics

Protective Device-Circuit Breaker Characteristics Usually calibrated for 40C. For specific shipboard application 50C calibration is also available		
Ratings	Description	Remarks
Continuous current rating (Ir)	a. Continuous current carrying capacity without tripping. This is also known as long-time pick-up.	
Long time delay	This is a pause in tripping time to allow temporary overload ride through. Can be adjusted from 3 to 25 seconds for around 6X current	
Short time pick-up	Controls the amount of high current the circuit breaker will remain closed against for short periods of time, allowing better coordination. Adjustable between 1.5 to 10 times the continuous current setting of the circuit breaker.	
Short time delay	Amount of time (from 0.05 second to 0.2 second in fixed time or 0.2 second at 6 times I rated in the $I^2 t$ ramp mode) a circuit breaker will remain closed against current in the pick up range. This function is used with the short time pick-up function to achieve selectivity and coordination. (A predetermined override automatically preempts the setting at 10.5 times the maximum continuous current setting.	
Instantaneous pick-up	Current level at which the circuit breaker trips without any intentional delay. The instantaneous pick-up function is adjustable from 2 to 40 times the continuous current rating of the circuit breaker.	

A short-circuit condition occurs when two or more energized conductors come in contact with each other or when there is a failure of conductor insulation. For shipboard power systems, the *three-phase short-circuit* condition also known as bolted fault causes the peak value of the fault current.

The short-circuit occurs when a very low impedance path is created through which the full generator or system energy is applied. The resultant current is usually several times greater than the normal circuit current. If protection devices do not detect the situation and consequently the faulted circuit is not opened the resultant current can cause extensive damage to the equipment and system.

Similarly if the device that opens the circuit is not designed to break the high current, several things may happen, all of which are detrimental to safety. Some of the scenarios are as follows:

a. The energy released by the high fault current may cause the device to explode.

b. The contacts opening the circuit may weld together and allow the resultant energy to pass through the system.

c. The contacts may start to open; the current may reestablish itself across the contact gap causing an arc flash. The arc may ignite other material and lead to a fire.

It is important to determine the value of the potential current likely to occur under short-circuit conditions and ensure that the devices installed to interrupt the fault current are rated to withstand it and operate in a programmed manner.

Similarly, it is important to ensure that devices connected in the short-circuit current path, for example motor starters, are adequately sized to withstand the short-circuit current for the duration required by the protective devices to open.

All of these requirements are managed under the short-circuit and coordination of protective devices. This fault current coordination and management is required to maintain a healthy electrical system.

2.12.1 Fault Current Calculation Fundamentals

The inception of electrical fault is a power system can not be predicted. Estimation of short-circuit path impedance can be very complex. In addition, although the generator or system power may be known, other consumers on the system append to that energy under short-circuit conditions.

Table 2.12   Load Flow and Voltage Drop Data Table

Balanced Load Flow and Voltage Drop Data Summary for Ship Service SWBD-1							
Circuit#	Name	Type	Rating AMP	Cable Type	Cable Lengti I	VD%	Remarks
1P401	PNLFDR	FDR	245	XXX MCM	xxjoT	X%	
2P401		FDR	156	XXX MCM	XXX FT	X%	
3P401	FIRE PUMP	FDR	200	XXX MCM	XXX FT	X%	
4P401		FDR	220	XXX MCM	XXX FT	X%	
5P401	CRANE	FDR	240	XXX MCM	XXX FT	X%	
6P401	STEERING	FDR	150	XXX MCM	XXX FT	X%	

2.13 ADJUSTABLE DRIVE FUNDAMENTALS

2.13.1 Advantages of ASD for Shipboard Application

Adjustable speed drives are becoming more prevalent than all other motor controllers due to the fact that they ease speed control over a wide range of speeds. The ship propulsion with ASD has been gaining popularity over the direct drive. These are the advantages of an electric propulsion drive over a direct shaft drive:

- Anchor windlass, winch, crane, elevator application
- HVAC application
- UPS
- Other ship service loads as appropriate

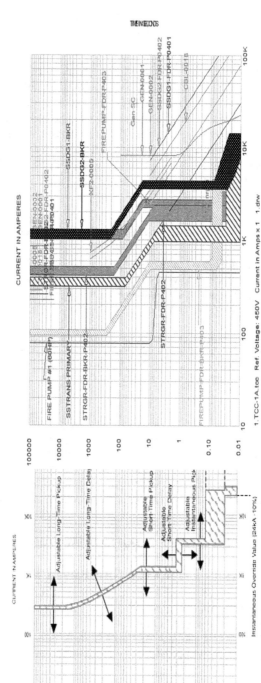

Figure 2.5 Typical Coordination of Protective Devices.

2.13.2 Disadvantages of VFD/ASD for Shipboard Application

The adjustable speed drive uses the fundamental shipboard power system with a 60 Hz, three-phase ungrounded source. It goes though a process to convert AC to DC (rectifier) and then DC to adjustable output (inverter). The carrier frequency for these conversion processes is at the range of multiples of the fundamental frequency, and range from 250 HZ to many kilohertz. These conversions generate many electrical noise issues contributing to transients and voltage overshoots. Electrical noise generates parasitic characteristics in the electrical system's behavior, including unacceptable harmonic distortion, and EMI and RFI issues. The electric propulsion drive manufacturer often addresses these issues with additional devices such as filters and transformers to minimize electric noise issues.

However, it is the responsibility of the system integrator to manage these electric noise matters as required to keep system-level distribution highly controlled.

The electric propulsion has become very common with ASD over direct drives. The disadvantages are:

- Harmonic currents that VFDs reflect back into the power system
- EMI and RFI feedback to the system

These are the cause of motor failure, cable failure, and protection system failure contributing to major arcing and three-phase fault in the system. For shipboard use of the ASD devices, it is the fundamental requirement to quantify types of electrical disturbances, manage those disturbances, and manage them at a safe level all the time. This requires major system-level design and development interventions so that there are no surprises after delivery of the ship.

The ship's operators must be familiar with the operational characteristics of ASD and the mitigation process for noise issues, and the ship must be provided with complete control, monitoring, and alarm systems.

In some cases, due to the criticality of the application, redundancy must be built into the system so that the graceful degradation of the system is accomplished automatically and manually to ensure the safety of the ship, system, and personnel.

2.14 FUNDAMENTALS OF ASD NOISE MANAGEMENT

Variable frequency solid-state devices
   Carrier frequency ranges:

- Solid state device carrier frequency – (Device turn on and turn off)
   1. SCRs (Silicon controlled rectifier) – What is the rise time of the variable speed drive's output IGBTs? (a) 2 kHz to 8 kHz can operate in the 250 to 500 Hz range (4 to 8 times fundamentals)

2. BJTs (Bipolar Junction Transistor) can operate in the 1 to 2 kHz range (16 to 32 times fundamentals)
3. Insulated gate bipolar transistors (IGBT) for the inverter section. IGBTs can turn on and off at a much higher frequency, 2 to 20 kHz (30 to 160 times fundamentals)
4. Typical AFE (Active front end with IGBT) switching frequency 3600 Hz (60 times fundamentals)

Solid-State Device Type	Carrier Frequency	Explanation
SCR	250 Hz to 500 Hz	4 to 8 times the fundamentals
BJT	1 kHz to 2 kHz	8 to 16 times the fundamentals
IGBT	2 Khz to 20 kHz	16 to 160 times the fundamentals
IGCT		

2.15 ELECTRICAL NOISE MANAGEMENT (SEE CHAPTER 7 FOR ADDITIONAL DETAILS)

- Low pass filter
- Band-pass filters
- Active filters
- Use of reactor
- Sinewave filters
- EMI and RFI filters
- High performance motor output filter
- System voltage? (a) 230 V, 450 V, 690 V, 4160 V, 6600 V.
- Solid state device carrier frequency: SCRs can operate in the 250 to 500 Hz range, BJTs can operate in the 1 to 2 kHz range, insulated gate bipolar transistors (IGBT) for the inverter section. IGBTs can turn on and off at a much higher frequency, up to 20 kHz. Typical AFE switching frequency 3600 Hz.
- What is the rise time of the variable speed drive's output IGBTs? (a) 2 kHz to 8 kHz).
- Type of load: (a) constant torque, (b) variable torque (c) constant horsepower? (a) all.
- What speed range is required? (a) 1200, 1800, and variable.
- Will the motor always be powered by the VSD or is across-the-line bypass required? (a) both.
- What is the motor's nameplate voltage and frequency rating?
- Motor rating: (a) NEMA MG-1, Class 30; (b) Class 31.

NEMA MG1 Part 30 is titled "Application Considerations for Constant Speed Motors Used on a Sinusoidal Bus with Harmonic Content and General Purpose Motors Used with Adjustable-Voltage or Adjustable-Frequency Controls or Both." NEMA MG1 Part 30.2.2.8 designates that for motors with a nominal insulation rating of equal to or greater than 600 V, the peak voltage rating of the insulation system has to be equal to or greater than 1 kV with a rise time of greater than or equal to 2 microseconds.

NEMA MG1 Part 31 is titled "Definite-Purpose Inverter-Fed Polyphase Motors." NEMA MG1 Part 31.4.4.2 designates that for the same nominal voltage rating as above the peak voltage rating of the insulation system needs to be 3.1 times V-rated with a rise time of equal to or greater than 0.1 microseconds. There is some debate as to whether V-rated in a nominal 480 V system is 480 V or the motor nameplate line-to-line voltage rating of 460 V. The difference is relatively small, 1426 V vs. 1488 V.

- What is the horsepower or kW rating of the motor?
- Will the conductors between the motor and the AC drive be shielded or non-shielded? (a) Both
- What type of raceway will be used for the conductors between the VFD drive and the motor? (a) Tray type
- Line reactor: (a) limit inrush to rectifier, (b) rounding waveform, (c) reducing peak current, and (d) lowering harmonic current distortion such as (a) line reactor
- The sinwave filter converts the PWM waveform to a near sinusoidal waveform, allowing sensitive applications to take advantage of the efficiencies and savings that PWM output power supplies: SINWAVE filter
- High-frequency noise solution: EMI (Electro Magnetic Interference), RFI (Radio Frequency Interference), Common Mode Noise.
  Filters such as KRF use a combination of high-frequency inductors and capacitors to reduce noise in the critical 150 kHz to 30 MHz frequency range.
- Low Pass Filter: THD to 5%

Harmonic Guard Active Filter (HGA): By actively monitoring the load current, the HGA determines the proper current waveform for injection into the system to maintain an acceptable level of harmonic distortion (TDD reduction to 5% or better at full load).

Older VFD inverter designs typically used silicon controlled rectifiers (SCR) or bipolar junction transistors (BJT) as the switching components.

- SCRs can operate in the 250 to 500 Hz range.
- BJTs can operate in the 1 to 2 kHz range.
- Most modern VFDs use insulated gate bipolar transistors (IGBT) for the inverter section. IGBTs can turn on and off at a much higher frequency, up to 20 kHz. The higher carrier frequencies associated with IGBTs offer some key

advantages over older SCR and BJT inverters, but also have a severe trade-off to be discussed later.

IGBTs typically are not used in VFD rectifier front-ends. VFD rectifiers typically use SCRs or similar slower switching components. SCRs offer an advantage in that their simpler design is more robust, given variable input voltage quality, and they have a relatively low cost.

However, just as the higher carrier frequency of the IGBT on the inverter can cause issues, so can a lower frequency of SCRs on the rectifier front end. These slower switching frequencies on the front end can cause excessive harmonic distortion in the voltage source. Depending on the magnitude of the total harmonic distortion introduced by the VFD and other loads (lighting, computers, etc.) sharing the same service, mitigation may be required in accordance with IEEE 519-1992.

Some mitigation may be achieved by using a 12-pulse inverter instead of a 6-pulse, or adding line reactors or a phase shift zigzag-type transformer used for drive isolation. Drive isolation transformers are designed to protect the VFD from interference from upstream power disturbances. They do very little to mitigate the magnitude of harmonic currents that VFDs reflect back into the power system

2.16 MOTOR PROTECTION SOLUTIONS: DV/DT MOTOR PROTECTION OUTPUT FILTER

The drawback to the PWM scheme is that rapid switching transitions cause overshoots in voltage due to parasitic capacitance (also related to common mode current) and inductance in the motor's leads. The parasitic components behave according to the equations:

$$I = C \, dv/dt$$

and

$$V = L \, di/dt.$$

The faster the drive switches (or, as dv/dt rises) the higher the surge currents will be in the leads. This in turn causes high voltage pulses across the parasitic inductances.

In the end, the faster the pulses switch, the greater the impact of cable capacitance and inductance. These fast voltage pulses stress the motor's windings, causing heat and possibly premature failure.

There is also capacitance in the motor's bearings due to the grease and air preventing direct and continuous contact. Again $I = C \, dv/dt$ causes current to flow through the bearings, increasing as the switching speed rises. This can lead to premature bearing failure.

A motor-guard fault alarm indicates that the motor-guard is not providing adequate dv/dt and peak voltage protection. If the drive operation continues during a fault condition, the motor or other equipment may be damaged.

The output waveform dv/dt exceeds the rated maximum (1,000 V/μs for 480 V units and 1500 V/μs for 600 V units).

• The output waveform peak voltage exceeds the selected maximum (1000 V for 480 V units and 1500 V for 600 V units).

Remarks:

(a) There is widespread use of devices that are directed toward the management of electrical noise in the system. It is mandatory for any such equipment use to be properly analyzed so that equipment cost as well as the addition of the weight are properly considered as benefits for the system.

(b) Recent ARC flash related incidents may have been contributed to by rampant use of the devices without real pro and con analysis.

(c) Reactor units:

(d) Low Pass Filter: THD to 5%

(e) Harmonic Guard Active Filter:
    By actively monitoring the load current, the HGA determines the proper current waveform for injection into the system to maintain an acceptable level of harmonic distortion (TDD Reduction to 5% or better at Full Load)

(f) Motor Guard Sinewave Filter: Motor Protection Solutions
    The motor guard sinewave filter converts the PWM wave form to a near sinusoidal wave form, allowing sensitive applications to take advantage of the efficiencies and savings that PWM output power supplies and drives offer.
    Motor failures occur mainly due to one of two causes: insulation failure or bearing failure. These failures are mainly due to generated heat and/or excessive system voltage. VFD-operated motor failures due to heat and excessive voltage must be analyzed and properly mitigated.
    Motors are not 100 percent efficient, and require cooling. TEFC motors are cooled by a shaft-mounted fan. If a particular load application has a relatively high turndown ratio resulting in a very slow shaft speed, the cooling from the fan may be adversely affected. It cannot be assumed that a motor will accommodate an infinite turndown ratio without overheating its insulation system. Remember the "10 degree" rule of thumb discussed earlier. Motor manufacturers have recommended operation speed ranges for their motors. Operation restrictions on the turndown of the equipment connected to the motor should reflect these operation speed range recommendations.
    As stated earlier, IGBTs can switch on and off extremely fast. The speed with which they can switch from 0 V to full DC bus voltage is referred to as rise time, or dv/dt. There is a phenomenon called "reflected wave" that is exacerbated by the IGBT's characteristically fast rise time (around 0.1 microseconds). The situation occurs when there is a mismatch between the interconnecting cable impedance and the motor. The motor terminals reflect the voltage rise back on the cable. This reflection on longer cable lengths can reinforce subsequence pulses, resulting in increasing electrical resonance as

the carrier frequency is increased. This reflected wave can result in a transient voltage up to two times the DC bus voltage. Again, this DC bus voltage can be 1.414 times the AC input voltage. In a 480 V system, this can result in transients in excess of 1200 V. Faster rise times reduce the cable length at which this phenomenon is experienced. One manufacturer's general rule of thumb is that this can become an issue if the cable length between the VFD and the motor exceeds 15 ft. In real-world applications, having this short a length is rather ambitious. Other manufacturers have recommendations for the maximum acceptable carrier frequency. Another recommended solution is to provide filtering devices between the VFD and the motor to mitigate the voltage overshoot, but that also adds cost and complexity.

(g) Motor Protection Filter

Motor Protection Solutions

A motor-guard fault alarm indicates that the motor-guard is not providing adequate dv/dt protection and peak voltage protection. If the drive operation continues during a fault condition, the motor and equipment driven may be damaged.

Motor Protection Output Filter

The output waveform dv/dt exceeds the rated maximum (1,000 V/μs for 480 V units and 1500 V/μs for 600 V units).

• The output waveform peak voltage exceeds the selected maximum (1000 V for 480 V units and 1500 V for 600 V units).

Chapter 3

Power System Design, Development, and Verification

3.0 INTRODUCTION: DESIGN, DEVELOPMENT, AND VERIFICATION PROCESS

Shipboard power system design and development has become very complex due to high power generation, variable drive applications for propulsion, and other ship service requirements. The traditional design base is not sufficient to support the new challenges of high power generation, high power consumables such as electrical propulsion, and other adjustable drive applications. Traditional protection systems such as overload and short-circuit protection are not sufficient. This book addresses traditional design and development issues, namely:

(a) The system design process

(b) The system development process

(c) Verification of the design and development for operational requirements

(d) Verification of the design and development for regulatory requirements

(e) Verification of failure mode and effects analysis (FMEA) of the design (QFA, DVTP, and PSTP)

(f) Verification of system behavior and maintenance requirements for the operators

(g) Verification of training requirements for the design

(h) Verification of operators' readiness

3.1 TYPICAL DESIGN AND DEVELOPMENT OF POWER GENERATION AND DISTRIBUTION (SEE FIGURE 3.1)

Figure 3.1 is an example of a typical shipboard power generation and distribution system delivering power to a main lube oil pump. The 6600 V switchboard is

Shipboard Power Systems Design and Verification Fundamentals, First Edition. Mohammed M. Islam.
© 2018 the Institute of Electrical and Electronics Engineers, Inc. Published 2018 by John Wiley & Sons, Inc.

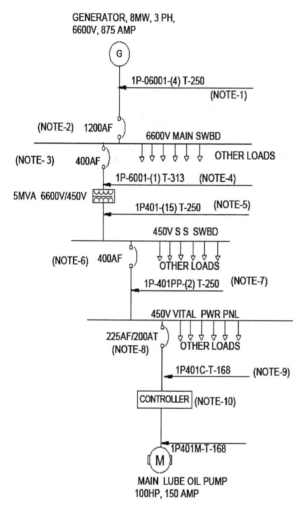

Figure 3.1   Typical 6600 V Power Generation and Distribution to a Main Lube Oil Pump Motor.

delivering power to the 480 V ship service switchboard through a 5 MVA, 6600 V to 450 V step down transformer. The 450 V switchboard is supplying power to a 450 V vital power panel. The vital power panel is supplying power to the main lube oil pump motor controller and the motor. For further clarification refer to Table 3.1.

3.2   FAILURE MODE AND EFFECT ANALYSIS (FMEA): DESIGN FUNDAMENTALS

3.2.1   Failure Mode and Effect Analysis (FMEA)

Due to the complex survivability requirements of various systems, the shipboard power system is required to be verified according to USCG FMEA guidelines. The following are the fundamentals of shipboard power system design.

Table 3.1 Example for 6600 V Power Generation to Supply Power to 450 V Main L.O. Pump (Refer to Figure 3.1)

Note#	Service	Explanation	Description	Remarks
1	Generator feeder: 1P-06001-(4) T-250.	Circuit designation 1P-06001 (4) three conductor 250 MCM cable.	(4) T-250, IEEE Std 45, 8 kV, 90 °C cable with ampacity rating 287 AMP per cable	6600 V, 875 A main generator feeder size Four cable is good for 4 × 287A = 1,148 AMP which provide a 31% margin. The voltage drop also needs to be addressed.
2	6.6 kV switchboard generator breaker	Generator circuit breaker 1250A frame with required tripping functions	1250 A frame. The trip settings are to be selected and set by the coordination of protective devices	
3	Circuit breaker for ship service 5 MVA transformer			3 MVA transformer 6.6 kV/450 V, 3-phase, 262 AMP primary side and/3849 AMP secondary side
4	Transformer primary feeder	Circuit designation 1P-6001 with (1) three conductor 313 MCM cable.	(1) T-313, IEEE Std 45, 8 kV, 90 °C cable with ampacity rating 298 AMP	
5	Transformer secondary feeder	Circuit designation 1P-401 with (15) three conductor 250 MCM cable.	(15) T-250, IEEE Std 45, 600 V, 90 °C cable with ampacity rating 259 AMP	
6	Power panel feeder breaker at the ship service switchboard for the load of 350 A		600 V, 400 A frame air breaker with 400 A thermal overload trip unit	
7	Vital power panel feeder size		Quantity 2, T-250, IEEE Std 45, 600 V, cable	The T-250 cable rating is 259 A
8	Circuit breaker for main lube oil pump		600 V, 225 A frame molded case breaker with 200 A thermal overload trip unit	Pump rating 100 HP, 150 A
9	Feeder for the main lube oil pump motor controller		Quantity 1, T-168, IEEE Std 45, 600 V, cable	IEEE Std 45 cable, T-168 is rated at 201 A
10	Motor controller overload		Thermal overload trip coil rating 170 A	

(a) Redundancy design intention
- Redundancy design intention and functional redundancy types
- Specification of subsystem or component groups

(b) Single failure propagation in redundant systems
- Failures, common causes, and systematic failure propagation
- Barriers and other compensating measures
- Failure propagation analysis at subsystem level

(c) Unit and subsystem FMEA
- Requirements to the unit FMEA including subsystem FMEA
- Allocation of unit requirements to subsystems/component groups
- Comparison of subsystem design intention with subsystem FMEA acceptance criterion

(d) FMEA of subsystems with redundancy

(e) FMEA report and compliance statement

The basic verification is outlined in the USCG regulations for failure mode and effect analysis (FMEA) supported by the qualitative failure analysis (QFA) and design verification test procedure (DVTP).

3.3 FAILURE MODE AND EFFECT ANALYSIS (FMEA) ELECTRIC PROPULSION SYSTEM DIESEL GENERATOR: DESIGN FUNDAMENTALS

3.3.1 Diesel Engine Operational Mode Selection

The diesel generator's (DG) operational features are vital for the electric propulsion system. The engine operation and generator operation must be understood by the operators for managing the electric power plant. The DG's operation is usually the automatic starting of the engine. The DG control system will have manual and automatic features. The manual control will always overtake the automatic controls. The selection for Automatic (AUTO) or Manual (MAN) is for the operators. In case of deviation between the selected setting and the position from local/remote selection, an alarm will sound and be displayed.

The DG's manual control, automatic control, safety management, and emergency shutdown function must be understood clearly to run the system and to take appropriate measures for the failure mode and effect analysis and operators' training.

Manual Functions

If MANUAL is selected with no applicable alarms in the system, the following commands will cause the effect:

- DG start command: DG start.
- DG stop command: DG stop. This function will unload the generator, and open the generator circuit breaker.

- Breaker close command: DG start, if not running, breaker sync.
- Breaker open command: DG unload, breaker open, DG cooling and stopping.

Automatic Functions

If AUTO is selected with no applicable alarms in the system, the following automatic actions may occur under the power management system (PMS):

- DG start/stop if automatic start/stop.
- Breaker synchronization is automatic.
- Load-sharing and Frequency control is automatic.

3.3.2 Diesel Generator Safety System Functions

(a) **The mimic for the generators will display the following information:**
Each generator's watt and VAR loading.
- Stator winding temperatures.
- Generator drive end and non-drive end bearing temperatures.
- Generator cooling air temperature.
- Generator cooler leakage alarms.
- Generator excitation failure alarm.
- Generator earth fault alarm.
- Generator frequency
- Generator voltage
- Generator current
- AVR monitoring (as provided)

(b) **Operation:**
- The reason for the DG's shutdown will be indicated in the status panel of the generator's mimic page element.

(c) **Alarm Functions:**
DGs can exhibit several levels of fault mode. The PMS will act to treat the conditions according to the severity of the alarm condition.
- Alarm 1: Start standby engine: Sick engine still running with priority to stop
- Alarm 2: Start standby engine and disconnect sick engine: Cooling and stopping of the sick engine.
- Alarm 3: Start standby engine and disconnect sick engine: Direct stop.
- Alarm 4: Sick switchboard: Start standby engine and disconnect connected generators before closing breaker of standby generator by direct in (no synchronization). Note: This only applies to the case of two DGs connected to one of the switchboards.
- Alarm 5: Delayed shutdown: Direct stop after a time delay.
- Alarm 6: Shutdown: Direct stop

3.3.3 Power Management Overview Mimic (Central Control Station and Switchboard)

The PMS Overview mimic is a simplified 6660 V power system single-line diagram with control blocks for each generator, as well as all generator and bus tie breakers connecting to the 6660 V Medium-Voltage Switchboard. Some of the information that follows is available from sub-mimic pages that pop up when selected from the main mimic page. Line current and power factor are displayed in tabular form. Other information available from this mimic page is as follows:

(a) Generator Incomer/Feeder/Bus Tie Circuit Breakers:
- Open/Closed/Earthed/Tripped by Fault/Ready indications.
- Local/Remote Control status
- Facility to Open/Close certain feeders via HMI (Central Control Station and Switchboard)

(b) Generators:
- Auto/Manual Selected.
- Active & Reactive Power (Watt & VAR) loading.
- Voltage & Frequency.
- Running/Stopped/Fault indications.
- Spinning reserve (Watt & VAR)
- Minimum number of generators to run
- Generator auto-start matrix
- Generator auto-stop matrix
- Generator operating mode (normal/fixed target load share/derated)

(c) Bus Ties:
- Current Flow/Power Flow

(d) Bus Sections:
- Voltage & Frequency.
- Protection Operated/Bus Dead/Control Supply Failed indications.

The operator can perform the following tasks from this mimic: (Central Control Station and Switchboard)

- Start/stop generators
- Derate generator output (Watt)
- Enable/disable frequency control
- Enable/disable active power (Watt) load sharing
- Set a generator to active power (Watt) fixed target load sharing mode (unbalanced operation)
- Enable/disable automatic (load dependent) start of generators.
- Enable/disable automatic (load dependent) stop of generators.
- Modify generator start/stop priority order.

- Enable/disable load shed functions.
- Enable/disable black start.
- View blackout data.
- View short-circuit alarms.

The following alarms are displayed on the HMI: (Central Control Station and Switchboard)

- Busbar Voltage High/Low
- Generator Voltage High/Low
- Busbar Frequency High/Low
- Generator Current High

3.3.4 Power Distribution Mimic Page

This mimic provides information on shipboard power distribution from the 450 V Ship Service Switchboard to the service loads in an active single-line diagram. This mimic page will display the following information:
450 V Incomer/450 V Bus Tie Circuit Breakers:
Open/Closed indications.

- Local/Remote control status
- Facility to Open/Close 450 V incomer circuit breakers via HMI (Central Control Station and Switchboard)

(a) 450 V Bus Sections:
- Earth Fault/Bus Dead indications.
- Emergency Board Incomer/Bus Tie Circuit Breakers:
- Open/Closed indications.
- Load shedding enabled/disabled
- Access to Load shedding schedule page

3.4 DESIGN VERIFICATION: GENERAL

A design is to meet the fundamentals of the requirements by contract specification, which is presented in an electrical one-line diagram such as Figure 3.1. There are design and development verification requirements. The basic verification is outlined in the USCG regulations called failure mode and effect analysis (FMEA) supported by the qualitative failure analysis (QFA) and design verification test procedure (DVTP).

3.4.1 Qualitative Failure Analysis (QFA)

The QFA must be prepared assuming the vessel is in its normal condition of operation and reflect the level of automation and manning level of the machinery plant, e.g., vessel underway in pilothouse control, all main engines in remote automatic operation, machinery space manned or unattended (depending on the vessel's manning level), and automatic power management system, if provided, active.

Checking the QFA's Failure Effects: a. Propulsion Control Systems. 46 CFR 62.35-5(e)(3).

3.4.2 Qualitative Failure Analysis (QFA) Basics

a. Perform this maintenance procedure after reviewing the results from the most recent infrared thermographic inspection of converter cubicles

b. Secure and tag out the converter as follows, in accordance with reference (1):
   i. Verify that the converter is in DRIVE SECURED mode.
   ii. Set the converter transformer feeder breaker local/remote control switch to the LOCAL position at the switchboard.
   iii. Verify that the converter transformer feeder breaker is in the OPEN position, and tag out the breaker in accordance with reference (1). Rack out the breaker.
   iv. Set the Exciter Transformer (ETF) feeder contactor control selector switch in the local (HV) position and verify the contactor is in OPEN position. Tag out in accordance with reference (1). Rack out the ETF contactor.

3.4.3 Process Failure Mode and Effect Analysis (FMEA): General

The QFA must be prepared assuming the vessel is in its normal condition of operation and reflect the level of automation and manning level of the machinery plant, e.g., vessel underway in pilothouse control, all main engines in remote automatic operation, machinery space manned or unattended (depending on the vessel's manning level), and automatic power management system, if provided, active.

Checking the QFA's Failure Effects: a. Propulsion Control Systems. 46 CFR 62.35-5(e)(3).

(1) Failures of the remote propulsion control system should be failsafe, such that the preset speed and direction (as-is) of thrust is maintained, until local manual or alternate manual control is in operation, or the manual safety trip (shutdown) is activated. This is required specifically for vessels with a single propulsion plant or single propeller.

(2) For a vessel with multiple and independently controlled propellers, a failure of one propulsion control system need not follow the preceding failsafe requirements. The failsafe options available in this case are:
   (i) Force both control systems to fail "as-is." Systems respond like a vessel with a single propulsion plant.
   (ii) Fail "as-is" of just the affected control system, while maintaining full control of the unaffected propulsion system.

(iii) Fail to "zero" thrust or trip of the affected propulsion system, providing a partial reduction of the normal propulsion capability as a result of malfunction or failure. Reduced capability should not be below that necessary for the vessel to run ahead at 7 knots or half-speed of the vessel, whichever is less, and is adequate to maintain control of the ship.

3.4.4 Qualitative Failure Analysis (QFA)-1

Table 3.2   Design Verification Matrix QFA

Item	System	Subsystem	Failure point	Unit	Local diagnostic	Alarm point	Local/ remote	Effects	Remarks
1	Main supply switchboard	Voltage	Transient	Voltage	Meter	Hi-low	Local/ remote	Degrade over all operations	System level
1a		Harmonics							
1b		Frequency		Hz		Fluctuating	Local/ remote	Degrade over all operations	System level
2	Prop transformer	Winding temp		C-degree	Meter	Hi point	Local/ remote	Degrade over all operations	

3.4.5 Explanation of the Detail Design Using QFA

Table 3.3   Typical Qualitative Failure Analysis

System: Propulsion Plant				Main Propulsion Generator		
1	2	3	4	5 Failure detection	6 Corrective	8
No.	Function	Failure mode	Failure cause	detection	action	Failure evaluation
1	Generation	Short-circuit		Alarm	No	Max. 80% total propulsion
2	Generation	Overcurrent		Alarm	No	Max. 80% total propulsion
3	Generation	Reverse power		Alarm	No	Max. 80% total propulsion
4	Generation	Undervoltage	Actual value failure	Alarm	No	Max. 80% total propulsion
5	Generation	Overvoltage	Actual value failure	Alarm	No	Max. 80% total propulsion

3.4.6 Design Verification Test Procedure (DVTP): General

The Design Verification Test Procedure (DVTP) document is required to be "Approved" and retained aboard the vessel. Using the DVTP document, design verification testing is required to be performed immediately after the installation of the automated equipment or before the issuance of the initial Certificate of Inspection. Final approval of the DVTP document is contingent upon satisfactory completion of onboard design verification tests in the presence of the Coast Guard. See 46 CFR 61.40-1(c), and 62.30-10(a).

Applicable to self-propelled vessels of 500 gross tons and over that are certificated under subchapters D, I, and U, and to self-propelled vessels of 100 gross tons and over that are certificated under subchapter H.

The Design Verification Test Procedure (DVTP) document is required to be "Approved" and retained aboard the vessel. Using the DVTP document, design verification testing is required to be performed immediately after the installation of the automated equipment or before the issuance of the initial Certificate of Inspection. Final approval of the DVTP document is contingent upon satisfactory completion of onboard design verification tests in the presence of the Coast Guard. See 46 CFR 61.40-1(c), and 62.30-10(a). Design verification testing is used to verify the automated vital system installations are designed, constructed and operate in accordance with the applicable requirements in 46 CFR Part 62. See 46 CFR 61.40-3.

The design verification test procedures may be incorporated with the qualitative failure analysis (QFA). See E2-18 Work Instruction. The DVTP document is a separate document from the Periodic Safety Test Procedure (PSTP) document. Both documents are required to be approved and retained aboard the vessel. See 46 CFR 61.40-1(c).

a. Title 46 CFR Parts 58, 61, and 62

b. Title 46 CFR Parts 111 and 112

c. Navigation and Inspection Circular (NVIC) 2-89, "Guide for Electrical Installations on Merchant Vessels and Mobile Offshore Drilling Units"

d. American Bureau of Shipping (ABS), "Rules for Building and Classing Vessels under 90 Meters in Length," 1996

e. Safety Of Life at Sea (SOLAS), Consolidated Editions, 1997, Chapter II-1, Part D

f. MSC Procedure E2-1, Vital System Automation Work Instruction

These guidelines were developed by the Marine Safety Center staff as an aid in the preparation and review of vessel plans and submissions. They were developed to supplement existing guidance. They are not intended to substitute or replace laws, regulations, or other official Coast Guard policy documents. The responsibility to demonstrate compliance with all applicable laws and regulations still rests with the plan submitter. The Coast Guard and the U. S. Department of Transportation expressly disclaim liability resulting from the use of this document.

If you have any questions or comments concerning this document, please contact the Marine Safety Center by e-mail or phone. Please refer to the Procedure Number: E2-18

- QFA General Acceptance Criteria:

Fail safe state must be evaluated for each subsystem, system or vessel to determine the least critical consequence. Lowest level of system component failure to be considered is: "easily replaceable component." 46 CFR 62.30-1(a).

The DVTP document, if submitted separately with the OFA document, must include the following QFA document information: Component Failures Considered are "Failure Effects" and "Failure Detection."

Examine the test instructions to insure that they closely or realistically simulate the failure of only the failed component of each of the failures considered in the failure analysis. For example: A PLC power supply module failure may be tested by removing the fuse to the power supply module, but a CPU failure (served by the same power supply module), should not be tested using the same power supply fuse, as it is desired the power supply to remain in operation, with just the CPU failing.

Test instructions should be prepared as if the vessel is underway, in pilothouse automatic pilothouse control, various machinery automation in normal underway mode of operation, and the engine room manned to the manning level design of the machinery plant.

Design verification testing using the failures considered in the QFA, the vital system automation installation, although supplied by various manufacturers, should function as an integrated system, i.e., various automated systems, although supplied by separate manufacturers, may be used to monitor the operational integrity of other systems and provide failure alarms.

3.4.7 Example-1: Propulsion Plant (DVTP) Design Verification Test Procedure

Programmable control or alarm system logic must not be altered after satisfactory completion of Design Verification Tests without the approval of the cognizant Officer in Charge, Marine Inspection. This comment should be included in the approval letter of the DVTP document to insure the cognizant OCMI and the ship's owner are aware of the requirements. See 46 CFR 62.25-25(a). This means that the DVTP document is only used during the initial issuance of the vessel's certificate of inspection or immediately after the installation of the automated equipment, and when the installed automated equipment is upgraded or altered. For the PSTP document, periodic safety testing is conducted at periodic intervals specified by the Coast Guard.

Table 3.4   Typical DVTP-1

Propulsion Plant (DVTP)			Main Propulsion Generator-Power System			
1	2	3	4	5	6	8
No.	Failure description	System effect due to Failure	Failure detection/ indication	Effect on overall ship performance due to failure	Corrective action	Failure evaluation (Remaining propulsion)
1	Generation	Short circuit		Alarm	No	Max. 80% propulsion
2	Generation	Overcurrent		Alarm	No	Max. 80% propulsion
3	Generation	Reverse power		Alarm	No	Max. 80% propulsion
4	Generation	Undervoltage	Actual value deviation	Alarm	No	Max. 80% propulsion
5	Generation	Overvoltage	Actual value deviation	Alarm	No	Max. 80% propulsion

3.5 SHIP SERVICE POWER SYSTEM DESIGN: SYSTEM-LEVEL FUNDAMENTALS (FIGURE 3.2)

(a) Two 6600 V generators

(b) Two 6600 V switchboards with bus tie

(c) Two propulsion motors, each connected to each section of the 6600 V switchboard

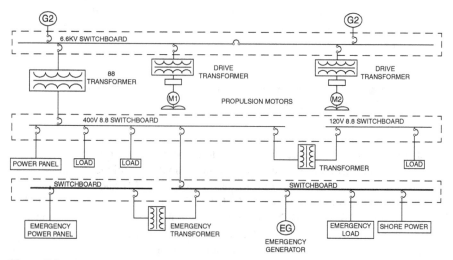

Figure 3.2   Typical Main Generators and Emergency Generator with Propulsion, Ship Service Distribution Switchboard, and Emergency Switchboard.

(d) Ship service transformer 6600 V to 450 V for 450 V ship service distribution switchboard

(e) 450 V to 120 V transformer for 120 V ship service distribution switchboard

(f) One 450 V emergency generator and 450 V emergency switchboard

(g) 450 V to 120 V emergency transformer for 120 V emergency switchboard

(h) 6600 V main switchboard
    Main generator breakers
    Ship service transformer circuit breaker

(i) 450 V Ship service switchboard
    Ship service transformer breakers
    Distribution load circuit breakers
    Emergency bus tie circuit breaker

(j) 450 V Emergency switchboard
    Emergency generator circuit breaker
    Ship service bus tie breaker section
    Emergency distribution circuit breakers

3.6 SINGLE SHAFT ELECTRIC PROPULSION (FIGURE 3.3)

There are four 6600 V generators, two 6.6 kV main switchboards and two ship service switchboards, one emergency generator, and one emergency switchboard. Redundant variable speed drives are provided to drive two winding propulsion motors on a single shaft. There are two thrusters, one connected to each 6.6 kV switchboard.

Each ship service switchboard is connected to two 6600 V main generators. Due to the number of generators, the load distribution and management of four ship service generators and one emergency generator can be complex. However, this design has proven to be very successful. The parallel operation and necessary protective control devices and their interlocking must however be clearly understood by the operator.

Port—Main switchboard—6600 V section

– Generator breakers

– Circuit breaker for propulsion transformer

– Circuit breaker for propulsion thruster

Starboard—Main switchboard—6600 V section

– Generator breakers

– Circuit breaker for propulsion transformer

– Circuit breaker for propulsion thruster

Port—Ship service switchboard—480 V section

– Motor controllers

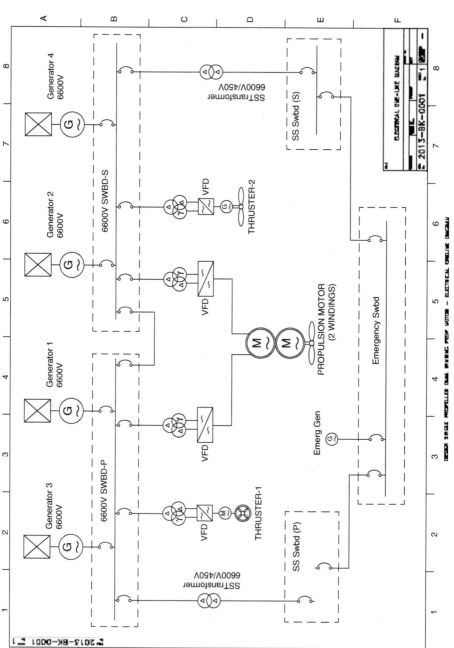

Figure 3.3 Typical 6600 V System Electrical One-Line Diagram for Propulsion with Single Shaft.

- Ship service load section for machinery auxiliaries, HVAC, deck machinery, and electrical services
- Shore power (open option of feeding from ship service or emergency switchboard)
- Emergency bus

Starboard—Ship service switchboard—480 V section

- Motor controllers
- Ship service load section for propulsion auxiliaries, HVAC, deck machinery, and electrical services
- Shore power (open option of feeding from ship service or emergency switchboard)
- Emergency bus tie

Starboard—Ship service switchboard—120 V section

- 115 V lighting
- Other 115 V services

Emergency switchboard—480 V section

- Emergency generator circuit breaker
- 480 V emergency loads
- Ship service bus tie breaker section

3.7 ELECTRICAL GENERATION AND DISTRIBUTION WITH DETAIL DESIGN INFORMATION (FIGURE 3.4)

In this example, there are four ship service generators, two ship service switchboards, one emergency generator, and one emergency switchboard. Each ship service switchboard is connected to two ship service generators. Due to the number of generators, the load distribution and management of four ship service generators and one emergency generator can be complex. The parallel operation and necessary protective control and their interlocking must however be clearly understood by the operator.

Port—Ship service switchboard—480 V section

- Generator breakers
- Motor controllers
- Ship service load section for machinery auxiliaries, HVAC, deck machinery, and electrical services
- Shore power (open option of feeding from ship service or emergency switchboard)
- Emergency bus tie

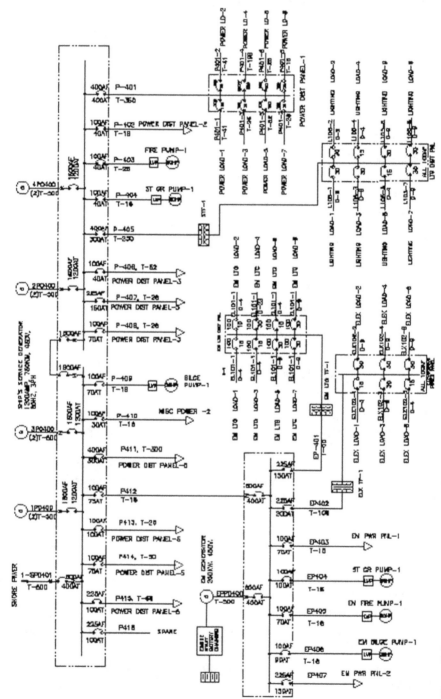

Figure 3.4 Detail Design of Low Voltage Distribution.

Starboard—Ship service switchboard—480 V section

- – Generator breakers
- – Motor controllers
- – Ship service load section for propulsion auxiliaries, HVAC, deck machinery, and electrical services
- – Shore power (open option of feeding from ship service or emergency switchboard)
- – Emergency bus tie

Starboard—Ship service switchboard—120 V section

- – 115 V lighting
- – Other 115 V services

3.8 ELECTRIC PROPULSION AND POWER CONVERSION UNIT FOR SHIP SERVICE DISTRIBUTION (FIGURE 3.5)

In this example, there are four 6600 V generators and two main switchboards providing power to two electric propulsion systems. The ship service switchboard is supplied with a transformer and two redundant motor generator sets. The motor generator sets are to ensure clean power at the ship service switchboard while the ship is under propulsion. However, the transformer is provided to provide power while the vessel is not under propulsion through the variable frequency drive. The variable frequency drive generates harmonics.

There is a dedicated ship service generator also to provide power to the ship service switchboard when the 6600 V system is turned off.

Port—Main switchboard—6600 V section

- – Generator breakers
- – Circuit breaker for propulsion transformer
- – Circuit breaker for propulsion thruster

Starboard—Main switchboard—6600 V section

- – Generator breakers
- – Circuit breaker for propulsion transformer
- – Circuit breaker for propulsion thruster

Ship service switchboard—480 V section

- – Generator breaker
- – Two motor generator circuit breakers
- – Motor controllers

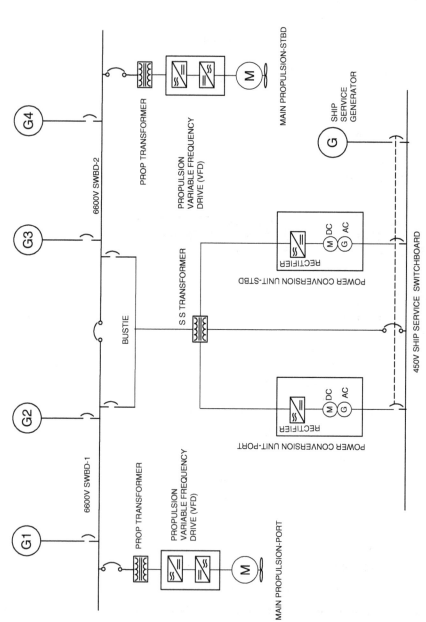

Figure 3.5 Detail Design of Medium Voltage Propulsion with MG Set for Clean Power.

- Ship service load section for machinery auxiliaries, HVAC, deck machinery, and electrical services
- Shore power (open option of feeding from ship service or emergency switchboard)
- Emergency bus tie

3.9 6600 V AND 690 V ADJUSTABLE SPEED APPLICATION WITH HIGH-RESISTANCE GROUNDING-1 (FIGURE 3.6)

There are four 6600 V generators, two main switchboards, two 690 V dedicated distribution switchboards, and 450 V ship service switchboards.

Port—Main switchboard—6600 V section

- Generator breakers
- Circuit breaker for propulsion transformer
- Circuit breaker for 690 V distribution system
- Circuit breakers for thrusters
- Generator neutral earthing resistors
- HGR, High resistance grounding system
- Circuit breaker for propulsion thruster

Starboard—Main switchboard—6600 V section

- Generator breakers
- Circuit breaker for propulsion transformer
- Circuit breaker for propulsion thruster
- Generator breakers
- Circuit breaker for propulsion transformer
- Circuit breaker for propulsion thruster

Port switchboard—690 V section

- Individual VFD
- Active front end VFD for multiple services
- HGR, High resistance grounding system

Stbd switchboard—690 V section

- Individual VFD
- Active front end VFD for multiple services
- HGR, High resistance grounding system

Port ship service switchboard—480 V section

Stbd ship service switchboard—480 V section

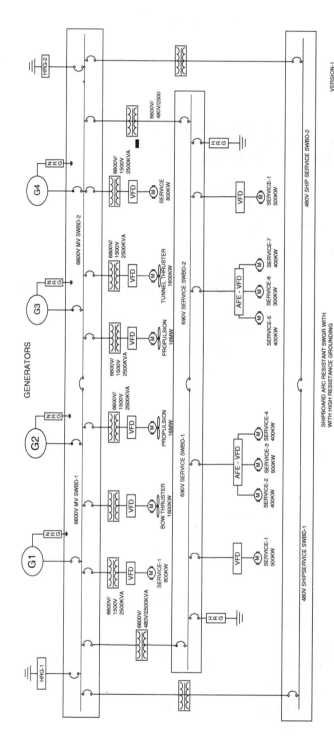

Figure 3.6 Typical Electrical One-Line Diagram for Shipboard MV and 690 V Adjustable Speed Application with (HRG) High Resistance Grounding-1.

3.10 MV AND 690 V ADJUSTABLE SPEED APPLICATION WITH HIGH-RESISTANCE GROUNDING (FIGURE 3.7)

There are four 6600 V generators, two main switchboards, two ship service switchboards, one emergency generator, and one emergency switchboard.

Port—Main switchboard—6600 V section

- Generator breakers
- Circuit breaker for propulsion transformer
- Circuit breaker for 690 V distribution system
- Circuit breakers for thrusters
- Generator neutral earthing resistors
- HGR, High resistance grounding system
- Circuit breaker for propulsion thruster
- Circuit breaker for ship service transformer

Starboard—Main switchboard—6600 V section

- Generator breakers
- Circuit breaker for propulsion transformer
- Circuit breaker for propulsion thruster
- Generator breakers
- Circuit breaker for propulsion transformer
- Circuit breaker for propulsion thruster
- Circuit breaker for ship service transformer

Port switchboard—690 V section

- Individual VFD
- Active front end VFD for multiple services
- HGR, high-resistance grounding system

Stbd switchboard—690 V section

- Individual VFD
- Active front end VFD for multiple services
- HGR, high-resistance grounding system

Port ship service switchboard—480 V section

- Active harmonic filter

Stbd ship service switchboard—480 V section

- Active harmonic filter

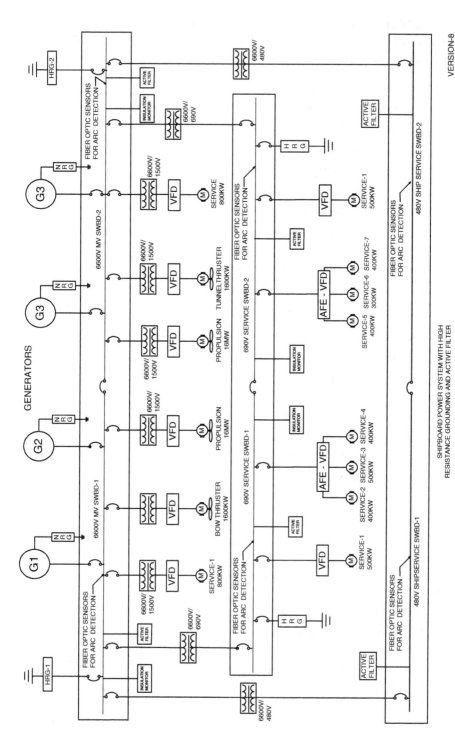

Figure 3.7 Typical electrical One-Line Diagram for Shipboard MV and 690 V Adjustable Speed Application with High-Resistance Grounding With Active Filter System.

3.11 FULLY INTEGRATED POWER SYSTEM DESIGN WITH ADJUSTABLE SPEED DRIVE (FIGURE 3.8)

There are four 6600 V generators, two main switchboards, two ship service switchboards, one emergency generator, and one emergency switchboard.

Port—Main switchboard—6600 V section

– Generator breakers
– Circuit breaker for propulsion transformer
– Circuit breaker for 690 V distribution system
– Circuit breakers for thrusters
– Generator neutral earthing resistor
– HGR, High resistance grounding system
– Circuit breaker for propulsion thruster
– Circuit breaker for ship service transformer

Starboard—Main switchboard—6600 V section

– Generator breakers
– Circuit breaker for propulsion transformer
– Circuit breaker for propulsion thruster
– Generator breakers
– Circuit breaker for propulsion transformer
– Circuit breaker for propulsion thruster
– Circuit breaker for ship service transformer

Port switchboard—690 V section

– Individual VFD
– Active front end VFD for multiple services
– HGR, high-resistance grounding system

Stbd switchboard—690 V section

– Individual VFD
– Active front end VFD for multiple services
– HGR, high-resistance grounding system

Port ship service switchboard—480 V section

– HGR, high-resistance grounding system

Stbd ship service switchboard—480 V section

– HGR, high-resistance grounding system

Emergency switchboard—480 V section

– HGR, high-resistance grounding system

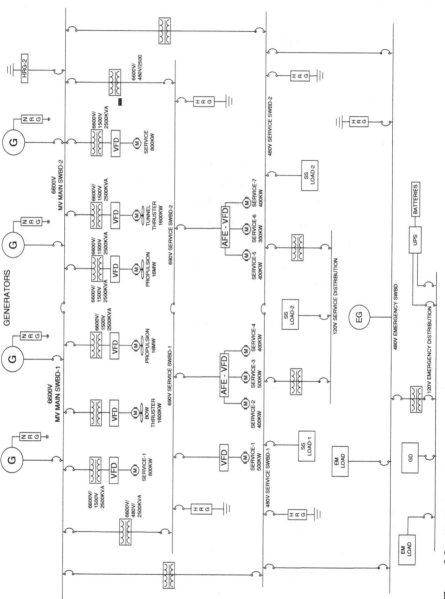

Figure 3.8 6600 V Generation, 690 V Distribution, 480 V Distribution, and 480 V Emergency Generation and Distribution.

3.12 VARIABLE FREQUENCY DRIVE (VFD) VOLTAGE RATINGS AND SYSTEM PROTECTION

Voltage Motor Insulation Stress when Driven by a Variable Frequency Drive

3.13 EXAMPLE 460 V, THREE-PHASE, FULL WAVE BRIDGE CIRCUIT FEEDING INTO A CAPACITIVE FILTER TO CREATE A 650 VDC POWER SUPPLY

(a) The peak voltage of a 460 V_{RMS} is:

$$VacPK = \sqrt{2} \times Vac_{RMS} = 1.414 \times 460 = 650 \, Vac_{PK}$$

Therefore, the DC output of the rectified 460 Vac is 650Vdc

(b) The peak voltage of a 690 V_{RMS} is:

$$Vac_{PK} = \sqrt{2} \times Vac_{RMS} = 1.414 \times 690 = 975 \, Vac_{PK}$$

Therefore, the DC output of the rectified 690Vac is therefore 975Vdc

3.14 SPECIAL CABLE AND CABLE TERMINATION REQUIREMENTS FOR VARIABLE FREQUENCY DRIVE APPLICATION

The variable frequency drive (VFD) application requires special VFD power cables. These VFD cables are much more complicated than the standard power cable. There are special requirements to terminate the drain wire and the shield to manage electric noise such as high frequency noise and EMI/RFI. The VFD cables are in general:

– Three power conductors
– Three drain wires
– Overall shield

3.15 HARMONIC MANAGEMENT REQUIREMENTS FOR VARIABLE FREQUENCY DRIVE APPLICATION

The variable frequency drive (VFD) application requires special attention to harmonic generation and harmonic management. In general the ship service distribution system related total harmonic distortion (THD) should not be more than 5%, as recommended by IEEE-45. In special bus system the THD can be 8%, Refer to IEEE-45-2002 clause 31.

Guidelines for shipboard VFD application and harmonic management:

– IEEE 519 was developed for shore-based power systems with diode-based six-pulse rectifiers. It is important to understand the use of IEEE 519 for shipboard application.

Table 3.5 VFD Related Voltage Ratings AC-DC

VAC RMS Voltage-Input	VAC Peak Voltage	Rectified VDC = VAC_{PK}	Inverted VAC_{PK}	VAC RMS Voltage-INV-Output	Voltage Swing-INV Out-Peak-Peak	VFD Cable Insulation
Rectifier Input-$VAC_{PK(RMS-IN)}$	Rectifier Input-VAC_{PK}	$VDC = VAC_{PK}$	$VDC = VAC_{PK}$ (Inverted)	$VDC = VAC_{RMS-OUT} = \frac{Vac\,pk}{\sqrt{2}}$	VAC_{PP}	System Ratings
120 V_{RMS-IN}	170 VAC_{PK}	170 VDC	340 VAC_{PK}	170 VDC	340 VAC_{PP}	600 V/1 KV
208 V_{RMS-IN}	294 VAC_{PK}	294 VDC	558 VAC_{PK}	294 VDC	558 VAC_{PP}	600 V/1 KV
460 V_{RMS-IN}	650 VAC_{PK}	650 VDC	1300 VAC_{PK}	650 VDC	1300 VAC_{PP}	2 KV
600 V_{RMS-IN}	849 VAC_{PK}	849 VDC	1698 VAC_{PK}	849 VDC	1698 VAC_{PP}	2 KV
690 V_{RMS-IN}	975 VAC_{PK}	975 VDC	1950 VAC_{PK}	975 VDC	1950 VAC_{PP}	2 KV
720 V_{RMS-IN}	1036 VAC_{PK}	1018 VDC	2036 VAC_{PK}	1018 VDC	2036 VAC_{PP}	2 KV
2000 V_{RMS-IN}	2828 VAC_{PK}	2828 VDC	5656 VAC_{PK}	2828 VDC	5656 VAC_{PP}	5 KV

- There are many different type of rectifiers which must be analyzed for harmonic management
- The PWM inverters generate harmonics at higher frequencies. The load side of the VFD generates harmonics and propagates throughout the power system. Therefore, VFD supplier recommendation should be followed for the selection of the VFD cable and termination of the VFD cable.
- These VFD cables are much more complicated than the standard power cable. There are special requirements to terminate the drain wire and the shield to manage electric noise such as high frequency noise and EMI/RFI. The VFD cable are in general:
- Three power conductors
- Three drain wires
- Overall shield
- Individual power cable shield (Option)

3.16 SWITCHGEAR BUS BAR AMPACITY, DIMENSION, AND SPACE REQUIREMENTS

3.16.1 Bus Bar Rating for English Dimensions (Inches)

Table 3.6   Bus Bar Ratings

	BAR Size in Inches	DC Ampacity Rating	60 HZ AC Ampacity Rating	Remarks
	Bare Copper BUS BAR Ampacity Table Based on 50 C Ambient			
ONE	3/4 × 1/8	250	250	
	1 × 1/8	330	330	
	1−1/2 × 1/8	500	500	
	1−1/2 × 3/16	580	570	
	2 × 3/16	760	745	
	1 × 1/4	490	480	
	1−1/2 × 1/4	685	675	
	2 × 1/4	920	900	
	3 × 1/4	1380	1280	
	4 × 1/4	1730	1700	
	5 × 1/4	2125	2000	
	6 × 1/4	2475	2300	
	8 × 1/4	3175	2875	
TWO	2 × 1/4	1525	1450	
	3 × 1/4	2225	2050	
	4 × 1/4	2800	2550	
	5 × 1/4	3100	2975	
	6 × 1/4	4000	3450	
	8 × 1/4	5100	4250	

Table 3.6   (*Cont.*)

	Bare Copper Bus Bar Ampacity Table Based on 50 C Ambient			
	Bar Size in Inches	DC Ampacity Rating	60 HZ AC Ampacity Rating	Remarks
THREE	3 × 1/4	3035	2550	
	4 × 1/4	3875	3225	
	5 × 1/4	4700	3880	
	6 × 1/4	5500	4400	
	8 × 1/4	6875	5300	
FOUR	3 × 1/4	3300	3050	
	4 × 1/4	4500	4250	
	5 × 1/4	5425	5000	
	6 × 1/4	6300	6000	
	8 × 1/4	7200	7100	

3.16.2 Bus Bar Rating for Metric Dimensions (Millimeter: MM)

Table 3.7   Bus Bar Ampacity Based On 45C Ambient

	Bare Copper BUS BAR Ampacity Table Based on 45 C Ambient				
BAR Size in MM (Width X Thickness)	60 HZ AC Ampacity Rating (One BAR)	60 HZ AC Ampacity Rating (2 in Parallel)	60 HZ AC Ampacity Rating (3 in Parallel)	60 HZ AC Ampacity Rating (4 in Parallel)	Remarks
15 × 3	230	390	470		
20 × 3	290	485	560		
20 × 5	395	690	900		
20 × 10	615	1145	1635		
25 × 3	355	580	650		
25 × 5	475	820	1040		
30 × 3	415	670	735		
30 × 5	555	940	1170		
30 × 10	835	1485	2070		
40 × 5	710	1180	1410		
40 × 10	1050	1820	2480	3195	
50 × 5	860	1410	1645	2490	
60 × 5	1020	1645	1870	2860	
60 × 10	1460	2430	3235	4075	
80 × 5	1320	2080	2265	3505	
80 × 10	1860	2985	3930	4870	
100 × 10	2240	3530	4610	5615	
120 × 10	2615	4060	5290	6360	

3.16.3 Nominal Working Space Requirements

Table 3.8   Typical Working Space Requirements

Normal Voltage to Ground (V)	Normal Voltage Phase to Phase (V)	Condition	Working Clearance	Reduced Clearance in Way of Stiffeners and Frames
0-150	120 V, 120 V/240 V 208 V-Y, 240 V DELTA	1	3 FT (0.91 m)	2 FT (0.61 m)
0-150	120 V, 120 V/240 V 208 V-Y, 240 V DELTA	2	3 FT (0.91 m)	2.5 FT (0.76 m)
0-150	120 V, 120 V/240 V 208 V-Y, 240 V DELTA	3	3 FT (0.91 m)	3 FT (0.91 m)
151-600	440 V Y OR DELTA 600 V Y OR DELTA	1	3 FT (0.91 m)	2.5 FT (0.76 m)
151-600	440 V Y OR DELTA 600 V Y OR DELTA	2	3.5 FT (1.07 m)	3 FT (0.91 m)
151-600	440 V Y OR DELTA 600 V Y OR DELTA	3	4 FT (1.22 m)	3.5 FT (1.07 m)
601-2500	2400V Y OR DELTA 4160 V Y OR DELTA	1	3 FT (0.91 m)	2.5 FT (0.76 m)
601-2500	2400V Y OR DELTA 4160 V Y OR DELTA	2	4 FT (1.22 m)	3.5 FT (1.07 m)
601-2500	2400V Y OR DELTA 4160 V Y OR DELTA	3	5 FT (1.52 m)	4 FT (1.22 m)
2501-9000	2400V Y OR DELTA 4160 V Y OR DELTA	1	4 FT (1.22 m)	3.5 FT (1.07 m)
2501-9000	2400V Y OR DELTA 4160 V Y OR DELTA	2	5 FT (1.52 m)	4 FT (1.22 m)
2501-9000	7200 V Y OR DELTA 13.8 kV Y OR DELTA	3	6 FT (1.82 m)	5 FT (1.52 m)
9001-25000	7200 V Y OR DELTA 13.8 kV Y OR DELTA	1	5 FT (1.52 m)	4 FT (1.22 m)
9001-25000	7200 V Y OR DELTA 13.8 kV Y OR DELTA	2	6 FT (1.82 m)	5 FT (1.52 m)
9001-25000	7200 V Y OR DELTA 13.8 kV Y OR DELTA	3	9 FT (2.74 m)	6 FT (1.82 m)

Where Conditions are as Follows:

Condition-1: Exposed live parts on one side and no live or grounded parts on the other side of the working space. Or exposed live parts on both sides effectively guarded by suitable insulation materials. Insulated wire or insulated bus bars operating at not over 300v shall not be considered live parts.

Condition-2: Exposed live parts on one side and grounded parts on the other side

Condition-3: Exposed live parts on both sides of the work space (not guarded as in Condition-1) with the operator between.

Exceptions: Working space shall not be required in back of assemblies such as dead front switchboards where there are no renewable or adjustable parts such as fuses or switches on the back and where all connections are accessible from other than back

---

The NEC Table 110.34 (A)

Exception: Working space shall not be required in back of equipment such as dead-front switchboards or control assemblies where there are no renewable or adjustable parts (such as fuses or switches) on the back and where all connections are accessible from locations other than the back.

Where rear access is required to work on nonelectrical parts on the back of enclosed equipment, a minimum working space of 762 mm (30 in.) horizontally shall be provided.

The provisions of 110.34 are conditional, just like the requirements in 110.26; that is, some of the requirements are applicable only where the equipment "is likely to require examination, adjustment, servicing, or maintenance while energized."

(A) Separation from Low-Voltage Equipment.

Where switches, cutouts, or other equipment operating at 600 volts, nominal, or less are installed in a vault, room, or enclosure where there are exposed live parts or exposed wiring operating at over 600 volts, nominal, the high-voltage equipment shall be effectively separated from the space occupied by the low-voltage equipment by a suitable partition.

Exception: Switches or other equipment operating at 600 volts, nominal, or less and serving only equipment within the high-voltage vault, room, or enclosure shall be permitted to be installed in the high-voltage vault, room, or enclosure.

3.17 MEECE (MANAGEMENT OF ELECTRICAL AND ELECTRONICS CONTROL EQUIPMENT) COURSE OUTLINE REQUIREMENTS: USCG

Any applicant successfully completing the 35-hour Marine Electric Propulsion course and presenting our certificate of training will be considered to have successfully demonstrated the competencies Operate main and auxiliary machinery and associated control systems of Table A-III/1 of the STCW Code; AND Start up and shut

Table 3.9 Meece Theoretical Course Summary

Meece-Manage Operation of Electrical and Electronics Control Equipment: Theoretical Course Summary				
Main Group	Subgroup	Subject	Duration Hours	Cumulative Time
1.0	1.0	Orientation	0.5	0.5
2.1		**Manage Operation of Electrical & Electronics Control Equipment-Theoretical (5 Hours)**		
2.1	2.1.1	Marine Electro-Technology	2.0	2.5
2.1	2.1.2	Electronics and Power Electronics	2.0	4.5
2.1	2.1.3	Automatic Control Engineering & Safety Devices	1.0	5.5
2.2		**Design Features & System Configuration of Automatic Control Equipment & Safety Devices (6 Hours)**		
2.2	2.2.1	Engine – General (All Engines) (Regulations)	1.0	6.5
2.2	2.2.2	Main Engines 2.0 (0.5 + 1.5)	0.5	7.0
2.2	2.2.2	Main Engines (1 hr 30 min)	1.5	1.5
2.2	2.2.3	Generation and Distribution	2.0	3.5
2.2	2.2.4	Steam Boiler	1.0	4.5
2.3		**Design Features & System Configuration of Operational Control Equipment for Electric Motor (15.5 Hours)**		
2.3	2.3.1	Three-phase induction motors	2.0	6.5
2.3	2.3.2	Three-phase synchronous motors	0.5	7.0
2.3	2.3.2	Three-phase synchronous motors	0.5	0.5
2.3	2.3.3	Effect of frequency and voltage on AC motor (continued)	1.5	1.5
2.3	2.3.4	Motor control & protection	2.0	4.0
2.3	2.3.5	IGBT motor control	1.0	5.0
2.3	2.3.6	Motor Speed control by thyristor	1.0	6.0
2.3	2.3.7	Three-phase generators	1.0	7.0
2.3	2.3.7	Three-phase generators	1.0	1.0
2.3	2.3.8	Three-phase transformer	2.0	3.0
2.3	2.3.9	Distribution	2.0	5.0
2.3	2.3.10	Emergency Power	1.0	6.0
2.4		**Design Features of HV Installations (4 Hours)**		
2.4	2.4.1	Design Features of HV Installations	1.0	7.0
2.4	2.4.1	Design Features of HV Installations	1.0	1.0
2.4	2.4.2	Operational safety of HV Installations	2.0	3.0
2.5		**Features of Pneumatic & Hydraulic Control Equipment (1.0 Hour)**		
2.5	2.5.1	Hydraulic Control equipment	0.5	3.5
2.5	2.5.2	Pneumatic Control equipment	0.5	4.0
2.6		**Unattended Machinery Space-ACCU (1.5 Hours)**		
2.6	2.6.1	Unattended Machinery Space-ACCU	1.5	5.5
2.7		**Dynamic Positioning System (30 Minutes)**		
2.7	2.7.1	Dynamic Positioning System	0.5	6.0
3.0	3.0	Test	1.0	7.0
		TOTAL HOURS = 35		

down main propulsion and auxiliary machinery, including associated systems; Operate, monitor and evaluate engine performance and capacity; Maintain safety of engine equipment, systems and services of Table A-III/2 of the STCW Code.

High Voltage Safety:

Any licensed engineer successfully completing the 35-hour High Voltage Safety course and presenting our certificate of training at a Regional Exam Center, will satisfy the:

(1) training and assessment requirements of the Seafarers' Training, Certification and Watch keeping Code, Section A-III/1, Tables A-III/1, Function: Maintenance and repair at the operational level, for the competency of Maintain marine engineering systems, including control systems; safety and emergency procedures; OR,

(2) training and assessment requirements of the Seafarers' Training, Certification and Watch keeping Code, Section A-III/2, Table A-III/2, Function: Maintenance and repair at the management level, for the competency Ensure safe working practices.

Chapter 4

Power Generation and Distribution

4.0 INTRODUCTION

Shipboard electrical power generation is generally for ship service power supported by emergency generators. If the propulsion system is also electrical then dedicated power generation is required for propulsion power. If the service requirement is for additional heavy loads then either the propulsion generator may support ship service loads or a dedicated ship service generator is required to support that load.

Due to the complexity of the service requirements, there are many options available for design engineers to consider. However, the fundamentals of basic electrical design remain the same, as the design must be safe and adequate to ensure that there is no generator overloading. Therefore, the verification of the adequacy of power generation is done by electric plant load analysis (EPLA). In addition, redundancy requirements must be complied with.

The ship's service power is generated by ship service generators, propulsion generators, emergency generators, and uninterruptible power sources. In the case of a prime mover-driven propulsion system, ship service electric power is generated by ship service generators. In the case of an electrical propulsion system, the ship's electrical propulsion power as well as ship service are generated by the propulsion generator, often supplemented by a smaller ship service generator. The regulatory body requirement is to provide redundant generators, ensuring the availability of electrical power under all operating conditions. It is also a requirement to run the generators under the most fuel-efficient conditions. The shipboard electrical loading scenarios typically are: at shore, at anchor, maneuvering, cruising, and other specific service-related loading. The generator's prime movers must be provided with the required starting system, governor system, and loading characteristics, ensuring the compatibility of the intended service. The number of generators and their ratings are to be properly calculated so that the electrical power generation is sufficient for all

Shipboard Power Systems Design and Verification Fundamentals, First Edition. Mohammed M. Islam.
© 2018 the Institute of Electrical and Electronics Engineers, Inc. Published 2018 by John Wiley & Sons, Inc.

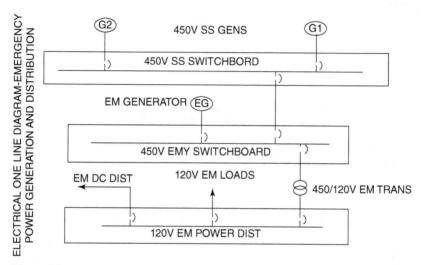

Figure 4.1   Typical Shipboard Power Generation and Distribution-1.

operating conditions of the ship. In order to ensure proper control and monitoring of these generators, IEEE 45 recommends a number of instruments in the switchboard as a minimum requirement. The propulsion generator ratings are higher than the ship service generator ratings as they provide propulsion load and ship service loads.

Shipboard power generation with N + 1 ship service power generation supported by emergency power generation is required by regulation. Figure 4.1 is an example ship service power with two generators and one emergency generator. The N + 1 interpretation for this design is that one ship service generator must be able to support all ship service loads under normal operational conditions. One emergency generator is to support emergency conditions only. The emergency generator must not be used for normal operational conditions but must be available as a precondition for ship operation. The emergency generator should be able to start automatically and provide power to the emergency loads if and when the ship service generator creates a blackout situation. Therefore, the pros and cons of design variations are explained in this book.

Figure 4.2 is an example of a sample shipboard power generation and distribution system with four main service generators and one emergency generator. The N + 1 interpretation for this design is that two to three main generators will support all loads under normal operational conditions. The emergency generator will power emergency loads when the running generator goes to the black-out situation. The emergency generator will not be used for normal operational conditions but must be available as a precondition for ship operation. The emergency generator should be able to start automatically and provide power to the emergency loads if and when the main generators running create a blackout situation.

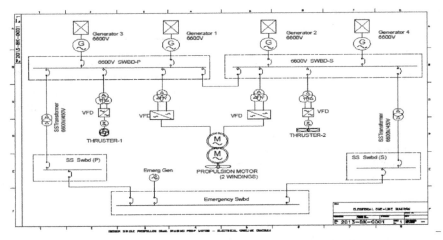

Figure 4.2   Typical All Electric Shipboard Power Generation and Distribution-2.

Figure 4.2 shows shipboard electrical design with electrical propulsion with variable frequency drive.

Table 4.1 explains the design scenarios of Figure 4.2.

4.1 GENERATION SYSTEM REQUIREMENTS

In accordance with Table 3 of IEEE 45-2002, the following voltages are recognized as standard.

The typical usage and frequencies associated with these voltages are briefly summarized below.

Generators

Modern shipbuilding, both military and commercial, employs mostly AC power generation and AC distribution. The generators are synchronous machines, with a magnetizing winding on the rotor carrying a DC current, and a three-phase stator winding where the magnetic field from the rotor current induces a three-phase sinusoidal voltage when the rotor is rotated by the prime mover. The frequency f [Hz] of the induced voltages is proportional to the rotational speed n [RPM] and the pole number p in the synchronous machine:

$$f = \frac{p}{2} \cdot \frac{n}{60}$$

For a 60 Hz system, a two-pole generator gives 3600 rpm, a four-pole at 1800 rpm, and a six-pole at 1200 rpm, etc. (US standard applications). For 50 Hz IEC applications,

Table 4.1   Text for Note 1, Use of Ship Service Transformer for MV Primary Generation and 480 V–120 V Ship Service Distribution

#	Description	Verification Remarks
1	This arrangement is the simplest and most economical design directly supplying power from the MV switchboard to the ship service switchboard. This traditional design complies with rules and regulations. However, in view of the VFD propulsion drive application, the simple step down transformer may not comply with the transient stability requirements as well as the total harmonic distortion (THD) limitation of the system. Therefore in the harmonic environment, other design features must be considered and added to comply with those requirements. For additional details refer to Note 2, Note 3, and Note 4.	It is mandatory to supply ship service type-1 power to certain equipment. Specific arrangement must be made to deliver that power.
2	The ship service power generation requirement is to provide dedicated ship service generators with redundant capabilities. As an alternate to the ship service generators MV transformer is used so that if the propulsion switchboard is energized, ship service power can be made available through the transformers.	If the power requirements cannot be met with the ship service generator design consideration is using 480 V, 3-phase output transformer. This application will require analytical verification of the main bus harmonic level. The regulatory body requirements of the harmonic level are given in Chapter 7, typically 5% or 5%–8% for specific applications. If the verification result shows the harmonic level above the acceptable THD limit then alternate arrangements must be made to limit harmonics. The harmonic management details are in chapter-7. If the design progresses without such verification during the preliminary design, the required calculation for the THD level will not be acceptable by the regulatory body. At that stage of the design, additional work must be done to satisfy regulatory requirements.
3		The design must be verified by the authority having jurisdiction before the design can be submitted to the approval authority.

Table 4.2 Standard Shipboard Voltages

Standard	AC(V)	DC(V)
Power utilization	115-200-220-230-350-440-460-575-660-2300-3150-4000-6000-10,600-13,200	115 and 230
Power generation	120-208-230-240-380-450-480-600-690-2400-3300-4160-6600-11,000-13,800	120 and 240

In synchronous generator design, the DC current is transferred to the magnetizing windings on the rotor by brushes and slip-rings. The new generators are equipped with a brushless excitation system. The brushless exciter generator is a synchronous machine with DC magnetization of the stator and rotating three-phase windings and a rotating diode rectifier.

The brushless exciter is typically mounted on the generator rotor shaft. The solid-state consists of rotating diodes, thyristors, and an RC snubber circuit. The permanent magnet generator (PMG) produces the excitation current for the generator. The synchronous generator excitation is controlled by an automatic voltage regulator (AVR) system that senses the terminal voltage of the generator and compares it with a reference value. According to most applicable regulations, the stationary voltage variation on the generator terminals should not exceed ± 2.5% of nominal voltage. For electrical generation system stability, transient load variation should exceed the voltage

Table 4.3 US and IEC Shipboard Power Generation and Distribution Levels at 50 HZ and 60 HZ

AC Voltage Generation 60 Hz and 50 Hz (V)	AC Voltage Distribution 60 Hz and 50 Hz (V)	Remarks
120	115	Mostly US applications
208	200	
230	220	Mostly IEC application and some US commercial applications
240	230	
380	350	
400	380	50 Hz
450	440	Mostly military applications
480	460	Mostly commercial applications
600	575	
690	660	Mostly IEC applications
2400	2300	50 Hz
3300	3150	60 Hz and 50 Hz
4160	4000	60 Hz and 50 Hz
6600	6000	50 Hz
11,000	10,600	60 Hz and 50 Hz
13,800	13,200	60 Hz and 50 Hz

variation of − 15% to + 20% of the nominal voltage. In order to maintain the transient voltage requirement, the AVR is normally also equipped with a feed-forward control function based on measuring the stator current.

In addition to the magnetizing winding, the synchronous generator rotor is also equipped with a damper winding, which consists of axial copper bars threaded through the outer periphery of the rotor poles, and short-circuited by a copper ring in both ends. The main purpose of the damper winding is to introduce an electromagnetic damping to the stator and rotor dynamics. A synchronous machine without a damper winding is inherently without damping and would give large oscillations in frequency and load sharing for any variation in the load.

The determination of power ratings for generators and prime movers requires careful calculation and analysis. The regulatory body requirement is to prepare a comprehensive electrical load analysis for each electrical power consuming load such as propulsion, propulsion auxiliaries, ship service auxiliaries, heating ventilation and air-conditioning (HVAC), normal lighting, emergency lighting onboard ship, and the load classification such as emergency service, vital service, non-vital service etc.

4.2 IEEE STD 45-2002, ABS-2002 AND IEC FOR GENERATOR SIZE AND RATING SELECTION

Ship service generator size, rating, and quantity requirements are very well defined by IEEE 45, ABS, and USCG with some slight differences. IEEE 45 requirements are in section 7, ABS-2002 requirements are in part 4, chapter 8, section 3, and USCG requirements are in 46 CFR, regulations subparts 111-10 and 111-12. The propulsion generator size, rating, and quantity requirements are somewhat different from ship service generators. It is very important to understand the operational requirements prior to adapting a set of rules. Due to the size of the propulsion generator, the quantity of generators and redundancy requirements may be different than for small size ship service generators. The propulsion generator requirements are given in IEEE 45, section 31. The most relevant generator size, rating, and quantity requirements are quoted as follows.

IEEE Std 45-2002, 7.4.2, Selection and Sizing of Generators (Extract)

In determining the number and capacities of generating sets to be provided for a vessel, careful consideration should be given to the normal and maximum load demand (i.e., load analysis) as well as for the safe and efficient operation of the vessel when at sea and at port. The vessel must have at least two generating sources. For ships, the number and rating of the main generating sets should be sufficient to provide one spare generating set (one set not in operation) at all times to service the essential and habitable loads. For MODUs, with the largest generator off-line, the combined capacity of the remaining generators must be sufficient to provide normal (non-drilling) load demands.

In selecting the capacity of an AC generating plant, particular attention should be given to the starting current of AC motors supplied by the system. With one generator held in reserve and with the remaining generator set(s) carrying the minimum load necessary for the safe operation of the ship, the voltage dip resulting from the starting current of the largest motor on the system should not cause any motor already running to stall or control equipment to drop out. It is recommended that this analysis be performed when total horsepower of the motor capable of being started simultaneously exceeds 20% of the generator nameplate kVA rating. The generator prime-mover rating may also need to be increased to be able to accelerate motor(s) to rated speed. Techniques such as soft starting (i.e., reduced voltage autotransformer starters, electronic soft starters, and variable frequency drives) may be utilized to reduce the required capacity of generators when motor starting is of concern.

Note: For SOLAS (Safety Of Life At Sea) requirement refer to SOLAS Chapter II-1, Regulation 41

IEEE Std 45-2002, Clause 31.3.1, General (electric propulsion prime movers) (extract)

The design of an integrated electric power system should consider the power required to support ship service loads and propulsion loads under a variety of operating conditions, with optimum usage of the installed and running generator sets.

In order to prevent excessive torsional stresses and vibrations, careful consideration should be given to coordination of the mass constants, elasticity constants, and electrical characteristics of the system. The entire system includes prime movers, generators, converters, exciters, motors, foundations, slip-couplings, gearing, shafting, and propellers... .

Systems having two or more propulsion generators, two or more propulsion drives, or two or more motors on one propeller shaft should be so arranged that any unit may be taken out of service and disconnected electrically, without affecting the other unit.

IEEE Std 45-2002, Clause 31.4, Prime movers for integrated power and propulsion plants

Prime movers, such as diesel engines, gas turbines, or steam turbines, for the generators in integrated electric power systems shall be capable of starting under dead ship conditions [see dead ship definition by ABS that follows] in accordance with requirements of the authority having jurisdiction. Where the speed control of the propeller requires speed variation of the prime mover, the governor should be provided with means for local manual control as well as for remote control.

The prime mover rated power, in conjunction with its overload and the large block load acceptance capabilities, should be adequate to supply the power needed during transitional changes in operating conditions of the electrical equipment due to maneuvering, sea, and weather conditions. Special attention should be paid to the correct application of diesel engines equipped with exhaust gas-driven turbochargers to ensure that sudden load application does not result in a momentary speed reduction in excess of limits specified in Table 4.

When maneuvering from full propeller speed ahead to full propeller speed astern with the ship making full way ahead, the prime mover should be capable of absorbing a proportion of the regenerated power without tripping from over-speed when the propulsion converter is of a regenerative type. Determination of the regenerated power capability of the prime mover should be coordinated with the propulsion drive system. The setting of the over-speed trip device should automatically shut down the unit when the speed exceeds the designed maximum service speed by more than 15%. The amount of the regenerated power to be absorbed should be agreed to by the electrical and mechanical machinery manufacturers to prevent over-speeding.

Electronic governors controlling the speed of a propulsion unit should have a backup mechanical fly-ball governor actuator. The mechanical governor should automatically assume control of the engine in the event of electronic governor failure. Alternatively, consideration would be given to a system, in which the electronic governors would have two power supplies, one of which should be a battery. Upon failure of the normal supply, the governor should be automatically transferred to the alternative battery power supply. An audible and visual alarm should be provided in the main machinery control area to indicate that the governor has transferred to the battery supply. The alternative battery supply should be arranged for trickle charge to ensure that the battery is always in a fully charged state. An audible and visual alarm should be provided to indicate the loss of power to the trickle charging circuit. Each governor should be protected separately so that a failure in one governor will not cause failure in other governors. The normal electronic governor power supply should be derived from the generator output power or the excitation permanent magnet alternator. The prime mover should also have a separate over-speed device to prevent runaway upon governor failure.

4.3 ABS-2002 SECTION 4-8-2-3.1.3 GENERATOR ENGINE STARTING FROM DEAD SHIP CONDITION (EXTRACT)

Dead ship (blackout) condition is the condition under which the main propulsion plant, boiler and auxiliaries are not in operation due to the unavailability of power from the main power source. See ABS 4-8-2-4-1-1/7.7. In restoring the propulsion, no stored energy for starting the propulsion plant, the main source of electrical power and other essential auxiliaries is to be assumed available. It is assumed that means are available to start the emergency generator at all times. The emergency source of electrical power may be used to restore the propulsion, provided its capacity either alone or combined with other source of electrical power is sufficient to provide at the same time those services required to be supplied by ABS 4-8-2/5.5.1 to 4-8-2/5.5.8.

The emergency generator and other means needed to restore the propulsion are to have a capacity such that the necessary propulsion starting energy is available within 30 min. of dead ship (blackout) condition as defined above. Emergency generator stored starting energy is not to be directly used for starting the propulsion plant, the main source of electrical power and/or other essential auxiliaries (emergency generator excluded). For steam ships, the 30 min. time limit is to be taken as the time from dead ship (blackout) condition to light-off of the first boiler.

Note: For SOLAS requirement refer to SOLAS Chapter II-1, Regulation 42.3.4 and 43.3.4

IEEE Std 45-2002, Clause 7.3.4, Governors (propulsion and ship service generator engines)

The prime-mover governor performance is critical to satisfactory electric power generation in terms of constant frequency, response to load changes, and the ability to operate in parallel with other generators.

The steady state speed variation should not exceed 5% (e.g., 3 Hz for a 60-Hz machine) of rated speed at any load condition.

Each prime mover should be under control of a governor capable of limiting the speed, when full load is suddenly removed, to a maximum of 110% of the rated speed. It is recommended that the speed variation be limited to 5% or less of the over-speed trip setting.

The prime mover and regulating governor should also limit the momentary speed variation to the values indicated in this sub-clause. The speed should return within 1% of the final steady state speed in a maximum of 5 seconds or as set by the limits specified in Table 7 IEEE Std 45-2002.

For emergency generators, the prime mover and regulating governor shall be capable of assuming the sum total of all emergency loads upon closure to the emergency bus. The response time and speed deviation shall be within the tolerances indicated in the Table 7 IEEE Std 45-2002.

IEEE 45-2002 Table 7— Response time and speed deviation requirements

Load (%)	Response time (s)	Speed deviation (%)
0 to 50, 50 to 0	5.0	10
50 to 100, 100 to 50	5.0	10

Generator sets should be capable of operating successfully in parallel when defined as follows: If at any load between 50% and 100% of the sum of the rated loads on all generators, the load (kW) on the largest generator does not differ from the other by more than ± 15% of the rated output or + 25% of the rated output of any individual generator, whichever is less, from its proportionate share. The starting point for the determination of the successful load distribution requirements is to be at 75% load with each generator carrying its proportionate load.

At least one voltage regulator should be provided for each generator. Voltage regulation should be automatic and should function under steady state load conditions between 0% and 100% load at all power factors that can occur in normal use. Voltage regulators should be capable of maintaining the voltage within the range of 97.5% to 102.5% of the rated voltage. A means of adjustment should be provided for the voltage regulator circuit. Voltage regulators should be capable of withstanding shipboard conditions and should be designed to be unaffected by normal machinery space vibration.

Solid state voltage regulators are recommended for high reliability, long life, fast response, and stable regulation. Regulator systems should be protected from under-frequency conditions. It is recommended that voltage regulators for machines rated in excess of 150 kW be provided with under-frequency and over-voltage sensors for protection of the voltage regulators.

Under motor starting or short-circuit conditions, the generator and voltage regulator together with the prime mover and excitation system should be capable of maintaining short-circuit current of such magnitude and duration as required to properly actuate the associated electrical protective devices. This shall be achieved with a value of than [*sic*] not less than 300% of generator full-load current for a duration of 2 seconds, or of such additional magnitude and duration as required to properly actuate the associated protective devices.

For single-generator operation (no reactive droop compensation), the steady state voltage for any increasing or decreasing load between zero and full load at rated power factor under steady state operation should not vary at any point more than ± 2.5% of rated generator voltage. For multiple units in parallel, a means should be provided to automatically and proportionately divide the reactive power between the units in operation.

Under transient conditions, when the generator is driven at rated speed at its rated voltage, and is subjected to a sudden change of symmetrical load within the limits of specified current and power factor, the voltage should not fall below 80% nor exceed 120% of the rated voltage. The voltage should then be restored to within ± 2.5% of the rated voltage in not more than 1.5 s.

In the absence of precise information concerning the maximum values of the sudden loads, the following conditions should be assumed: 150% of rated current with a power factor of between 0.4 lagging and zero to be applied with the generator running at no-load, and then removed after steady state conditions have been reached.

For two or more generators with reactive droop compensation, the reactive droop compensation should be adjusted for a voltage droop of no more than 4% of rated voltage for a generator. The system performance should then be such that the average curve drawn through a plot of the steady state voltage vs. load for any increasing or decreasing load between zero and full load at rated power factor, droops no more than 4% of rated voltage. No recorded point varies more than ± 1% of rated generator voltage from the average curve.

Isochronous operation of a single generator operating alone is acceptable. However, where two or more generators are arranged to operate in parallel, it is recommended that isochronous kilowatt load sharing governors and voltage regulation with reactive differential compensation capabilities be provided. Care should be taken if operating machines in parallel to ensure that the system minimum load does not decrease and cause a reverse power condition.

If voltage regulators for two or more generators are installed in the switchboard and located in the same section, a physical barrier should be installed to isolate the regulators and their auxiliary devices.

Where power electronic devices (such as variable frequency drives, soft starters, and switching power supplies) create measurable waveform distortions (harmonics), means should be taken to avoid malfunction of the voltage regulator, e.g., by conditioning of measurement inputs by means of effective passive filters.

Power supplies and voltage sensing leads for voltage regulators should be taken from the "generator side" of the generator circuit breaker. Normally, voltage sensing leads should not be protected by an over-current protection device. If short-circuit protection is provided for the voltage sensing leads, this short-circuit protection should be set at no less than 500% of the transformer rating or interconnecting wiring ampacity, whichever is less. It is recommended that a means be provided to disconnect the voltage regulator from its source of power.

4.4 ADDITIONAL DETAILS OF SIZING SHIP SERVICE GENERATORS

Detailed electrical load analysis should be made to select the size of ship service generators. The radial electrical system generator sizing requirement is different from the ring distribution. The radial system in general consists of ship service generators and emergency generators. The ring bus distribution system often consists of multiple generators and may not have a dedicated emergency generator.

(a) Ship Service Generator for radial distribution: (for typical radial bus power distribution)
   – The generator size must be equal to or bigger than the maximum steady state worst-case load requirement of the system.
   – The generator must be able to sustain the largest direct on line (DOL) motor starting load (starting current and transient voltage dip) of the system. If the DOL starting requirement leads to bigger sizes of generator and engine, detailed calculations must be performed to replace the DOL starting system with wye-delta starting, with open transition or closed transition, auto-transformer starting, soft starting, or variable frequency drive. The calculations must be accompanied by trade study for service requirements, space requirements, and cost impact. Always remember, the radial distribution system may not have an additional generator to support the loading requirements.
   – The generator must be able to sustain the harmonic distortion effect of the system contributed by nonlinear loads in the system.

(b) Ship Service Generator for ring bus distribution system (for typical ring bus power distribution)
   – The ring bus distribution is usually set up with multiple generators in the system so that generators can be connected in parallel to support steady state loading under automatic or manual power management systems. In this configuration, the sizing of generators is not as complex as in the radial configuration. However, in the radial configuration, there is no dedicated emergency generator, where the redundant power feeder requirement can get complicated.

4.4.1 Engine Governor Characteristics

Governors for engine generator sets used in parallel operation should be of the electronic load sharing type. Such governors are specifically designed for isochronous operation at any load from zero to 100% load. These governors cause their respective prime movers or engines to share load proportional to their horsepower or kilowatt rating and as a function of direct measurement of the electrical load (in kilowatts) on a specific engine governor set. Proportional division is accomplished via feedback through a load share loop that connects all governors to all generator sets on-line at any given time. Such governors provide the following functions:

(a) Rapid response to load changes

(b) Stable system operation

(c) Paralleling dissimilar sized engine generator sets

(d) Speed regulation of .25%

4.4.2 Generator Voltage Regulator Characteristics

Due to sudden load application on the generator, the load current surge causes voltage change in the output voltage. This voltage drop is due to an internal voltage drop in the generator winding. This is called voltage dip in the system. Similarly during large load shedding the system produces overvoltage. An automatic voltage regulator is used to control rapid voltage changes in the system.

An automatic voltage regulator (AVR) can control the voltage fluctuation within (plus) (minus) 2.5% or better of the set value over the full load range. This AVR voltage regulation is called steady state voltage regulation. The transient voltage dip is usually limited to 15% for a specified sudden load change with recovery back to rated voltage within 1.5 seconds or better as required by application and acceptable to regulatory body.

In some applications for large loads where unusually large voltage surge is expected the AVR performance limit may be adjusted to the satisfaction of the customer. However, this adjustment should not cause system-level instability. If this adjustment cannot be managed, the large load, such as large cranes, application must be simulated to establish an acceptable voltage transient. However, this transient must be within an acceptable envelope.

4.4.3 How AVR works:

It is the function of the voltage regulator to control the excitation of the generator so as to maintain constant generator terminal voltage, within defined limits, e.g., + 6%, to − 10%. Since most engine generator sets used in today's power systems are brushless type generators, relatively inexpensive solid-state voltage regulators can be furnished to provide steady state regulation of 1% to 2% under any load condition from zero to 100% load. These solid-state controls are also capable of rapid response to load changes and of boost excitation to provide current magnitude capable of achieving selective coordination with the over-current protective devices in the power system. As with the governors, voltage regulators are also available with droop characteristics. The purpose of a droop characteristic in a voltage regulator is to enable the generator to share the reactive component of the load in proportion to the kilovoltampere (kVA) rating of the generator.

The amount of reactive load is also an important design factor. For example, if an engine is delivering 800 kW at .8pf, then the engine is delivering 1000 kVA. The reactive power, kilovar, equals the square root of kilovoltampere squared minus kilowatt squared. So for this example, the generator would be delivering 600 kvar reactive power. When generator sets are operating in parallel, allowing voltage to drop or droop, as the current out of the machine increases, it causes the generators to share load almost proportionately. Why "almost proportionately"? It stems from dissimilarities in current-transformer-to-full-load-rating ratio discrepancies. Because the electrical loads driven by today's power systems require precise frequency and voltage levels, droop compensation in the voltage levels, droop compensation in the voltage control for kilovar sharing is as objectionable as speed droop is for kilowatt sharing between the engines.

In most cases, the same regulator that would be used for droop compensation can also be used for crosscurrent compensation. The only difference lies in the connections of the regulator in the power system. Crosscurrent compensation for proportionate reactive load division (kilovar sharing) is a highly desirable design feature for paralleling systems employing automatic unattended operation. The designer should also specify crosscurrent compensation for the voltage regulation of these systems.

4.4.4 Droop Characteristics: Generator Set

When selecting engine generator sets for parallel operation, the size of the sets is determined by analyzing the voltage and frequency requirements in terms of the load, transient response, stability, and droop.

Droop is a function of the difference between no-load and full-load operation. It is the percent difference in the values based on the no-load value. For instance, for a

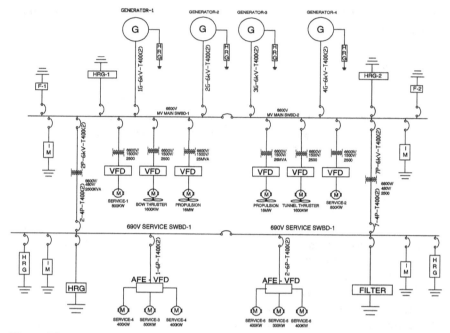

Figure 4.3   Ship Service Electric Generator for 460 V Generation and 450 V Distribution typical.

machine with 3% droop and a no-load frequency of 61.8 Hz, the full load frequency will be 97% of the no-load frequency of 60 Hz. The frequency droop for this setup is therefore 1.8 Hz. Similarly, if the voltage droop is 5% on a nominal system voltage of 480 V AC, the voltage droop is 24 V AC.

However, today's engine generator sets can be furnished with relatively simple and reliable electronic governors and voltage regulators that make these sets suitable for unattended automatic paralleling and load sharing. This is a basic key to paralleling engine generators for emergency power systems. The electronic governor provides isochronous operation, and automatic proportionate load division, which make possible the automatic paralleling of dissimilar size sets. The electronic governor provides a more adaptable engine generator set because it will permit paralleling at any time without necessitating adjustment or requiring droop. Similarly, voltage regulators are available to achieve automatic reactive load division to provide constant voltage systems. These devices help to make automatic unattended emergency power paralleling systems highly practical.

4.5 TYPICAL GENERATOR PRIME MOVER

These drawings show typical generator prime mover setups. These drawings are presented with minimum explanation and not intended to show a complete or proven design.

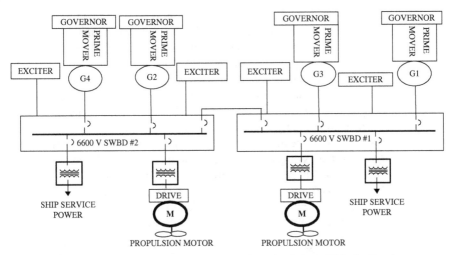

Figure 4.4   Electrical One-Line Diagram, Electrical Propulsion System (R2 Redundancy).

Electric propulsion system (ABS R2 redundancy)

The figures 4.3 & 4.4 shows an electric propulsion configuration with four prime movers, two high-voltage switchboards, and two sets of propulsion motors with their drives. This configuration is in compliance with ABS R2-S notation for redundant propulsion systems. In this configuration there is no requirement of physical separation by watertight and fireproof bulkhead in the machinery space, which means four engines and propulsion drives can be in the same machinery space.

Electric propulsion configuration with four prime movers. Two high-voltage switchboards, and two sets of propulsion motors with their drives. This configuration is an extension of Figure 4, with the additional requirement to meet ABS R2-S + notation, which means there is a separation between two machinery spaces with a watertight centerline bulkhead.

4.6 GENERATOR: TYPICAL PURCHASE SPECIFICATION (TYPICAL ELECTRICAL PROPULSION SYSTEM)

Generator quantity and size and usually established during concept design. However, the sizes of the generators are verified again by the following methods:

(a) The generator size is verified by developing electrical plant load analysis. The load analysis is required to be approved by the owner and then by a regulatory body. The generator is usually purchased as a generator set, such as the prime mover, and the generator. The generator set is weight critical for the ship. Therefore, size and volume must be optimized.

(b) The generator size usually consists of the requirement as established by the electrical plant load analysis (EPLA). There may be a design margin, which is for the design and development margin (D&D margin). This is to ensure additional capacity, which is used for the detail design and development cycle to account for missing loads. Again the D&D margin dictates the maturity of the design. For an electrical propulsion system with many megawatts of power generation requirement D&D can add substantial load leading to the next size of engine and generator. The D&D margin is usually around 5% at the early phase of the detail design. There may be a requirement for reserve margin for future growth. This reserve margin is an established design system deliverable. The reserve margin can be up to 20%, which also contributes to the engine and generator sizes. At the final stage there is a process to precisely match the engine and generator due to the fact that the EPLA calculated generator size may not exactly match the engine size.

(c) The number of generators usually follows the N + 1 rule ensuring one standby all the time. If the EPLA establishes a three-generator requirement, then three plus one for a total four generators.

(d) The generators shall be capable of operating in parallel and can load share equally.

(e) The sub-transient reactance of the generators shall be selected to limit short-circuit current such that with all generators connected to the switchboard bus plus all motor contributions, the interrupting rating of the circuit breaker shall not be exceeded during any system short-circuit. This feature is determined by short-circuit analysis, which is also a deliverable to regulatory body.

Chapter 5

Emergency Power System Design and Development

5.0 INTRODUCTION

(a) Emergency Generation and Distribution

USCG 46 CFR Subpart 112 outlines requirements for emergency power generation and distribution as a "Final Emergency Power Source." Some of the USCG requirements are extracted to provide better understanding of sizing an emergency generator and for providing emergency loads. In addition there are special requirements for interlocking the control system for systematic monitoring of bus voltage, initiating the engine start sequence, taking on emergency loads, and finally, when the main power is restored, redirecting all emergency loads to the ship's service power. For ship design, this requirement must be met without exception. There will be verification to establish that the requirements are met.

(b) Emergency No-Break Power Generation and Distribution

A black-out related to electrical power loss situation is not acceptable for shipboard control system, computerized system and for special applications. Therefore, the no-break power requirement is introduced. No-break power can consist of many different types of storage power. The battery-supported uninterruptable power supply with converter is shown, supplying 450 V three-phase power and 120 V three-phase power. The load demand may lead to very large battery systems, which may require additional UPS ventilation and other interlocking requirements.

Shipboard Power Systems Design and Verification Fundamentals, First Edition. Mohammed M. Islam.
© 2018 the Institute of Electrical and Electronics Engineers, Inc. Published 2018 by John Wiley & Sons, Inc.

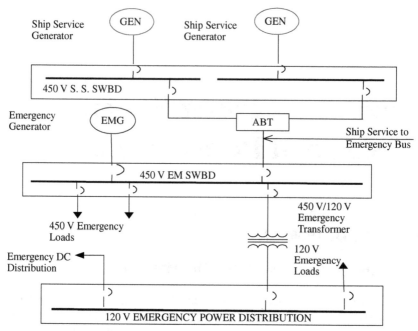

Figure 5.1   Typical Emergency Generator Distribution with Ship Service Power Generation.

5.1 USCG 46 CFR REQUIREMENTS: 112.05 (EXTRACT ONLY)

Purpose: Ensure a dependable, independent, and dedicated emergency power source with sufficient capacity to supply those services that are necessary for the safety of the passengers, crew, and other persons in an emergency and those additional loads that may be authorized.

Ensure main-emergency bus-tie disconnects automatically upon loss of potential at the emergency switchboard. Emergency generator starts automatically and provides required emergency power.

List of Emergency Loads (USCG CFR 46)

By regulation, the following emergency loads must be fed from the emergency switchboard. For emergency generator rating calculation the connecting loads must be calculated for 100 percent load factor: (Refer to USCG 46 CFR Chapter 1 Subpart 112.15 for details)

– Emergency fire pump (at least one fire pump with all auxiliaries must be fed from the emergency switchboard)

- Steering gear (for 1 steering gear system the power must be provided from ship service as well as emergency through automatic bus transfer switch)
- Watertight door closure (all)
- Emergency lighting
- Navigation and communication system—Emergency power supply with UPS backup
- Emergency power supply for the automation with UPS support
- Emergency bilge (1)

5.2 IEEE STD 45-2002, CLAUSE 6.1, GENERAL (EXTRACT)

Every vessel should be provided with a self-contained emergency source of electric power, generally a diesel-engine generator, gas turbine-driven generator, or storage batteries. These emergency sources of power and an emergency switchboard should be located in a space separate and remote from the main switchboard, that is above the uppermost continuous deck, aft of the collision bulkhead, outside the main machinery compartment, and readily accessible from the open deck. The emergency switchboard should be located in the same space as the emergency power source, in an adjacent space, or as close as practical.

IEEE Std 45-2002, Clause 6.2, Emergency Generators (Extract)

Each emergency generator should be equipped with starting devices with an energy storage capability of at least six consecutive starts. A single source of stored energy, with the capacity for six starts, should be protected to preclude depletion by the automatic starting system, or a second source of energy should be provided for an additional three starts within 30 min. If, after three attempts, the generator set has failed to start, an audible and visual "start failure" alarm should be activated in the main machinery space control station and on the navigation bridge. The starting sequence should be automatically locked-out until an operator can initiate the final three starting attempts from the emergency generator space.

IEEE Std 45-2002, Clause 6.4, Emergency Power Distribution System (Extract)

The emergency switchboard should be supplied during normal operation from the main switchboard by an interconnecting feeder. This interconnecting feeder should be protected against short-circuit and overload at the main switchboard and, where arranged for feedback, protected for short-circuit at the emergency switchboard. The interconnecting feeder should be disconnected automatically at the emergency switchboard upon failure of the main source of electrical power. Means shall be provided to prevent auto closing of the emergency generator circuit breaker should a fault occur on the emergency switchboard . . . .

Upon interruption of normal power, the prime mover driving the emergency power source should start automatically. When the voltage of the emergency source reaches 85% to 95% of nominal value, the emergency loads should transfer automatically to the emergency power source. The transfer to emergency power should be accomplished within 45 seconds after failure of the normal power source. If the system is arranged for automatic retransfer, the return to normal supply should be accomplished when the available voltage is 85% to 95% of the nominal value and the expiration of an appropriate time delay. The emergency generator should continue to run without load until shut down either manually or automatically by use of a timing device . . . .

The emergency switchboard should be arranged to prevent parallel operation of an emergency power source with any other source of electrical power (i.e. main power), except where suitable means are taken for safeguarding independent emergency operation under all circumstances. This will allow for the emergency generator to be used to supply non-emergency loads. This is useful when exercising and testing the emergency generator(s).

5.3 EMERGENCY SOURCE OF ELECTRICAL POWER: ABS 2010, 5.1.1 REQUIREMENT

Basic Requirement. A self-contained emergency source of electrical power is to be provided so that in the event of the failure of the main source of electrical power, the emergency source of power will become available to supply power to services that are essential for safety in an emergency. Passenger vessels are subject to the requirements in *5/13.5 of the ABS Guide for Building and Classing Passenger Vessels.*

ABS-2002, 4-8-2, 5.3.3, Separation from Other Spaces

Spaces containing the emergency sources of electrical power are to be separated from spaces other than machinery space of category A by fire rated bulkheads and decks in accordance with Part 3, Chapter 4 of ABS Rules or Chapter II-2 of SOLAS.

5.4 ABS EMERGENCY GENERATOR STARTING REQUIREMENT (ABS RULE FOR PASSENGER VESSELS)

Section 13.5.8: Starting Arrangements for Emergency Generator

Sets *13.5.8(a) General.* The emergency generator is to be capable of being readily started in their cold condition at a temperature of 0°C (32°F). If this is impracticable or if lower temperatures are likely to be encountered, heating arrangements are to be fitted.

13.5.8(b) Number of Starts. Each emergency generator, arranged to be automatically started, is to be equipped with starting devices with a stored energy capability of at least three consecutive starts. The source of stored energy is to be protected to preclude critical depletion (i.e., not to be depleted beyond a level where starting by manual intervention is still possible) by the automatic starting system, unless a second independent means of starting is provided. In addition, another source of energy is to be provided, for an additional three starts within thirty minutes unless manual starting can be demonstrated to be effective.

13.5.8(c) Stored Energy for Starting. The stored energy for starting the emergency generator set is to be maintained at all times, as follows:

(i) Electrical and hydraulic starting systems are to be maintained from the emergency switchboard.

(ii) Compressed air starting systems may be maintained by the main or auxiliary compressed air receivers through a suitable non-return valve or by an emergency air compressor which, if electrically driven, is supplied from the emergency switchboard.

B. All of the starting, charging and energy storing devices are to be located in the emergency generator space. These devices are not to be used for any purpose other than the operation of the emergency generating set, however, this does not preclude the supply to the air receiver of the emergency generating set from the main or auxiliary compressed air system through the non-return valve fitted in the emergency generator space.

5.5 TYPICAL EMERGENCY GENERATION AND DISTRIBUTION SYSTEM

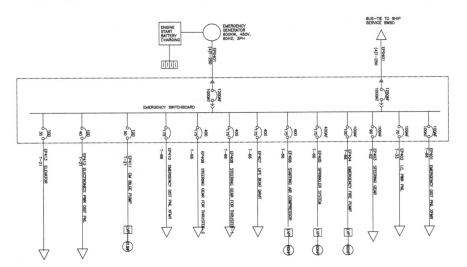

Figure 5.2   Typical Electrical One-Line Diagram for Emergency Distribution.

5.6 EMERGENCY GENERATOR AND EMERGENCY TRANSFORMER RATING: LOAD ANALYSIS (SAMPLE CALCULATION)

Table 5.1 is a sample electrical load analysis in support of selecting the emergency generator kilowatt rating and emergency transformer kVA rating. The load factor for an emergency load is required to be 100 percent by regulation.

5.7 EMERGENCY POWER GENERATION AND DISTRIBUTION WITH SHIP SERVICE POWER AND DISTRIBUTION SYSTEM

Shipboard emergency generation and distribution consist of a dedicated emergency switchboard and a dedicated emergency generator. The emergency switchboard is tied to the ship service switchboard through a bus tie. The emergency switchboard feeds the 120 V section of the emergency switchboard through 450 V/120 V transformer. The 450 V section of the emergency switchboard provides power to all 450 V emergency loads. The 120 V section of the emergency switchboard provides power to all 120 V loads.

The emergency switchboard is tied to two ship service switchboards through a bus tie and automatic bus transfer switch (ABT), refer to figures 5.1 and 5.2 for electrical oneline representation.

5.8 EMERGENCY TRANSFORMER 450 V/120 V (PER ABS)

In place of one three-winding transformer or three single-winding transformers in one enclosure with one primary breaker requirement, the ABS rules recommend the use of three separate single-phase transformers with dedicated circuit breakers, as shown in the figure.

NOTE:

1. Three single phase 450 V/120 V transformers, forming 120 V three-phase distribution. Each transformer is installed separately and each transformer is protected separately by a dedicated circuit breaker, refer to figure 5.3.

5.9 EMERGENCY GENERATOR STARTING BLOCK DIAGRAM

The starting system of the emergency generator engine is required to have redundant engine starting capabilities, ensuring that the engine is ready to get started, and take over emergency loads within 45 seconds of the failure of normal ship service power,

Table 5.1   Typical Electrical Load Analysis for Sizing Emergency Generator and Transformer

1	2	3	4	5	6	7	8
Item	Services	Qty	kW-Each	kW-Connected	Load factor	Total-kW Generator	Total-kW Transformer
1	Emergency Fire Pump System	1	150.0	150.0	1.0	150.0	
2	Steering Gear-1	1	98.0	98.0	1.0	98.0	
3	Steering Gear Thruster	2	40	80	1.0	80	
3	Watertight Door Closures	5	3.0	15.0	1.0	15.0	
4	Elevator	2	10	20	1.0	20.0	
5	Emergency Bilge	1	10	10	1.0	10.0	
6	Starting air compressor	1	10	10	1.0	10.0	
7	Emergency Lighting	All		70.0	1.0	70.0	70.0
8	Navigation & Communications System			15.0	1.0	15.0	15.0
9	Central Control System	1	15	15	1.0	15.0	15.0
10	UPS	2	10.0	20.0	1.0	20.0	20.0
11							
12	Grand Total—Generator					503.0	
13	Grand Total—Transformer (450/120 V)						120.0

NOTES

1. Emergency generator rating: (example only)
   - Emergency generator load analysis load factor is 100 (1.0) percent for all loads as shown in column 6
   - As shown in the calculation the generator output rating must be at least 503 kW.
   - The engine output must be 503/0.94 = 535 kW (Considering (94 percent) 0.94 generator efficiency)
   - The engine output rating should be equal to or greater than the calculated load 535 kW
   - Sometimes the engine increase is necessary to support a large motor starting transient.

2. Emergency transformer rating:
   - Emergency transformer rating in (kVA) = 120 kW/0.8 = 150 kVA (0.8 is the system power factor)
   - The transformer nameplate rating is 150 kVA

by detecting zero voltage at the switchboard bus. The starting systems are usually a combination of compressed air, hydraulic, and electrical UPS, see figure 5.4.

A typical starting system with one hydraulic and two electrical systems. The second electrical starting system is an additional capability to ensure redundancy if and when the criticality of the ship's performance is the shipowner's requirement.

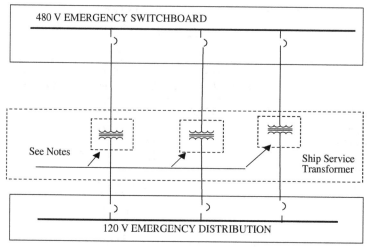

Figure 5.3   Emergency Transformer 480 V/120 V (Per ABS).

5.10 EMERGENCY GENERATION AND DISTRIBUTION DESIGN VERIFICATION

- – Emergency Fire Pump
- – Emergency Bilge Pump
- – Steering gear system
- – Emergency Lighting
- – Navigation Lights
- – Radio Communication

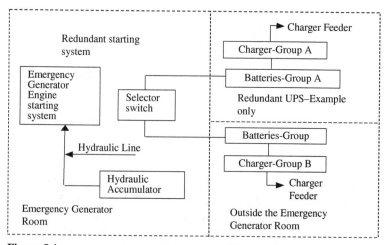

Figure 5.4   Emergency Generator Starting System Block Diagram.

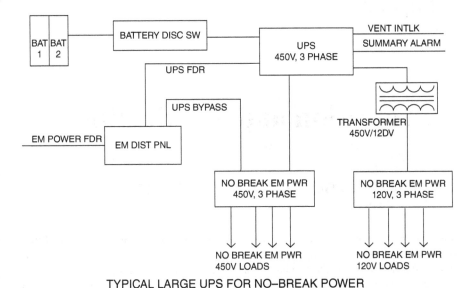

TYPICAL LARGE UPS FOR NO–BREAK POWER

Figure 5.5   No Break 450 V, 3-Phase Power Generation and Distribution.

- Internal Communication
- Navigational aids (i) Magnetic compass, (ii) Gyro compass, (iii) Radar, (iv) Echo-sounder, (v) Rudder angle indicator, and (vi) Propeller revolution counter
- Remote Propulsion Control and Monitoring System for ACC and ACCU Notations.

5.11 NO-BREAK EMERGENCY POWER DISTRIBUTION

(a) Battery bank with line disconnect

(b) Typical 450 V UPS, 450 V, 3-Phase. Due to the size of the UPS there may be forced ventilation and ventilation interlock.

(c) 450 V no-break distribution panel

(d) 120 V no-break distribution panel

Chapter 6

Protection and Verification

6.0 INTRODUCTION: PROTECTION SYSTEM FUNDAMENTALS

The shipboard power generation and distribution system is usually 60 Hz, three-phase, and ungrounded. A simplified shipboard electrical power generation and distribution system from generator to load is shown in Table 6.1. Every item in the table is identified uniquely in view of the service requirements. Each item will be discussed in view of the protection requirements. The design engineer must understand each item along with its attributes to make the design complete and to comply with rules and regulations.

The fault current analysis is performed to establish the mechanical strength and electrical properties of circuit breakers, switches, and bus bars. This equipment must be designed either to withstand the fault-generated energy, or the protective devices must be set to clear the fault promptly and safely.

The low-voltage shipboard power system utilizes air circuit breakers and molded case circuit breakers for overload as well as short-circuit protection. These devices are selected to detect unusual overcurrent in an electrical system and open the circuit if an overcurrent occurs. Electrical fault is one of the conditions that can cause an instantaneous extremely high current and analysis of the fault conditions is required to be well thought out when designing or modifying a shipboard electrical power system.

A short-circuit condition occurs when two or more energized conductors come in contact with each other or when there is a failure of the conductor insulation. For shipboard power systems, the *three-phase short-circuit* condition is also known as bolted fault.

At the instant a short-circuit occurs, a very low impedance path is created through which the full system energy is released. The resultant current is usually several times greater than the normal circuit current. If protection devices do not detect the short-circuit situation, the faulted circuit can cause extensive damage to the equipment and surroundings.

Shipboard Power Systems Design and Verification Fundamentals, First Edition. Mohammed M. Islam.
© 2018 the Institute of Electrical and Electronics Engineers, Inc. Published 2018 by John Wiley & Sons, Inc.

Table 6.1 Generator to Load Distribution and Protection

ITEM	DESIGNATION	DESIGN STEP	PROTECTION SEQUENCE	ONE LINE PRESENTATION
1	GENERATOR	DESIGN STEP-17	PROTECTION SEQUENCE-1	
2	GENERATOR FEEDER	DESIGN STEP-16	PROTECTION SEQUENCE-2	3 CONDUCTOR CABLE
3	GENERATOR SHORT CIRCUIT PROTECTION AND DISC MEANS	DESIGN STEP-15	PROTECTION SEQUENCE-3	SWBD #1 6600V 3PH 60HZ
4	TRANSFORMER SHORT CIRCUIT PROTECTION AND DISC MEANS	DESIGN STEP-14	PROTECTION SEQUENCE-4	
5	TRANSFORMER FEEDER	DESIGN STEP-13	PROTECTION SEQUENCE-5	3 CONDUCTOR CABLE
6	TRANSFORMER	DESIGN STEP-12	PROTECTION SEQUENCE-6	SERVICE TRANSFORMER 6600/450V
7	FEEDER TO SHIP SERVICE SWBD	DESIGN STEP-11	PROTECTION SEQUENCE-7	3 CONDUCTOR CABLE SHIP SERV SWBD
8	SHORT CIRCUIT PROTECTION AND DISC MEANS LOAD CENTER	DESIGN STEP-10	PROTECTION SEQUENCE-8	450V 3Ph
9	FEEDER TO VITAL LOAD CENTER	DESIGN STEP-9	PROTECTION SEQUENCE-9	3 CONDUCTOR CABLE
10		DESIGN STEP-8	PROTECTION SEQUENCE-10	VITAL LOAD CENTER
11	SHORT CIRCUIT PROTECTION AND DISC MEANS LOAD CENTER	DESIGN STEP-7	PROTECTION SEQUENCE-11	450V 3Ph
12	FEEDER TO VITAL POWER PANEL	DESIGN STEP-6	PROTECTION SEQUENCE-12	3 CONDUCTOR CABLE VITAL POWER PNL
13	SHORT CIRCUIT PROTECTION AND DISC MEANS LOAD CENTER	DESIGN STEP-5	PROTECTION SEQUENCE-13	450V 3Ph
14	FEEDER TO CONTROLLER	DESIGN STEP-4	PROTECTION SEQUENCE-14	3 CONDUCTOR CABLE
15	CONTROLLER WITH OVERLOAD PROTECTION	DESIGN STEP-3	PROTECTION SEQUENCE-15	CONTROLLER
16	MOTOR FEEDER	DESIGN STEP-2	PROTECTION SEQUENCE-16	3 CONDUCTOR CABLE
17	MOTOR	DESIGN STEP-1	PROTECTION SEQUENCE-17	MOTOR 450V 3Ph

Similarly, if the device that opens the circuit is not designed to break the high current, several things may happen, all of which are detrimental to safety. Some of the scenarios are as follows:

a. The energy released by the high fault current may cause the device to explode.

b. The contacts opening the circuit may weld together and allow the resultant energy to pass through the system.

c. The contacts may start to open; the current may reestablish itself across the contact gap, causing an arc flash. The arc may ignite other material and lead to a fire.

It is important to determine the value of the potential current likely to occur under short-circuit conditions and ensure that the devices installed to interrupt the fault current are rated to withstand it and operate in a programmed manner.

Similarly, it is important to ensure that devices connected in the short-circuit current path, for example motor starters, are adequately sized to withstand the short-circuit current for the duration required by the protective devices to open.

6.1 PROTECTIVE DEVICE: GLOSSARY

Breaking current – Value of prospective breaking current that a device is capable of breaking at a stated voltage under prescribed conditions of use and behavior.

Breaking time – Interval of time between the beginning of the opening time of a mechanical switching device and the end of the arcing time.

Circuit breaker frame – (1) The circuit breaker housing that contains the current carrying components, the current sensing components, and the tripping and operating mechanism. (2) That portion of an interchangeable trip molded case circuit breaker remaining when the interchangeable trip unit is removed. (100AF, 400AF, 800AF, 1600AF, etc.)

Continuous current rating (ampere rating) – The designated RMS alternating or direct current in amperes that a device or assembly will carry continuously in free air without tripping or exceeding temperature limits.

Drawout circuit breaker – An assembly of a circuit breaker and a supporting structure (cradle) so constructed that the circuit breaker is supported and can be moved to either the main circuit's connected or disconnected position without removing connections or mounting supports.

Electronic trip circuit breaker – A circuit breaker that uses current sensors and electronic circuitry to sense, measure and respond to current levels.

IDMT (Inverse Definite Minimum Time) – Time/current graded overcurrent protection. Basically, the more current put through the relay, the faster it goes.

Instantaneous pickup – The current level at which the circuit breaker will trip with no intentional time delay.

Instantaneous trip – A qualifying term indicating that no delay is purposely introduced in the tripping action of the circuit breaker during short-circuit conditions.

Insulated case circuit breaker (ICCB) – UL Standard 489 Listed non-fused molded case circuit breakers that utilize a two-step stored energy closing mechanism, electronic trip system, and drawout construction.

Interrupting rating – The highest current at rated voltage available at the incoming terminals of the circuit breaker. When the circuit breaker can be used at more than one voltage, the interrupting rating will be shown on the circuit breaker for each voltage level. The interrupting rating of a circuit breaker must be equal to or greater than the available short-circuit current at the point at which the circuit breaker is applied to the system.

Inverse time – A qualifying term indicating there is purposely introduced a delay in the tripping action of the circuit breaker, which delay decreases as the magnitude of the current increases.

KAIC – Kiloamperes interrupting capacity.

Long-time ampere rating – An adjustment that, in combination with the installed rating plug, establishes the continuous current rating of a full-function electronic trip circuit breaker.

Long-time delay – The length of time the circuit breaker will carry a sustained overcurrent (greater than the long-time pickup) before initiating a trip signal.

Long-time pickup – The current level at which the circuit breaker long-time delay function begins timing.

Making capacity (of a switching device) – Value of prospective making current that a switching device is capable of making at a stated voltage under prescribed conditions of use and behavior.

Molded case circuit breaker (MCCB) – A circuit breaker that is assembled as an integral unit in a supportive and enclosed housing of insulating material, generally 20 to 3000 A in size and used in systems up to 600 VAC and 500 VDC.

Opening time (of mechanical switching device) – Interval of time between the specified instant of initiation of the opening operation and the instant when the arcing contacts have separated in all poles. For circuit breakers: for a circuit breaker operating directly, the instant of initiation of the opening operation means the instant when the current increases to a degree big enough to cause the breaker to operate.

Peak let-through current – The maximum peak current flowing in a circuit during an overcurrent condition.

Short-circuit delay (STD) – The length of time the circuit breaker will carry a short-circuit (current greater than the short-circuit pickup) before initiating a trip signal.

Short-circuit making capacity – Making capacity for which prescribed conditions include a short-circuit at the terminals of the switching device.

Short-circuit breaking capacity – Breaking capacity for which prescribed conditions include a short-circuit at the terminals of the switching device.

Time Current Curve (TCC) – The method of ensuring selective coordination is to examine each overcurrent device's time current curve (TCC) and verify, for any value of current, that the protective device closest to the fault clears faster than any upstream device.

6.2 POWER SYSTEM PROTECTIONS

Quoted from the Marine Technology Society (MTS) guidelines for DP vessel design philosophy: Guidelines for MODU DP system and commercial ships (system protection functionality).

6.2.1 (MTS9.7.3:) Over current detection: This is the most basic form of protection and is applied at all levels in the power distribution systems for short circuit and over load protection. Over current can be detected by current transformers, fuses, magnetic over current or bi-metal strips with heating coils. At the main power distribution levels "protection-class" current transformers are used to provide digital relays with a signal representing the line current. Various current versus time curves are used to produce the required degree of coordination with other over current protection upstream and down stream.

Note: Protection class CTs may not provide the degree of accuracy required for instrument applications.

6.2.2 (MTS9.7.4:) Differential protection: Differential protection is a form of over current protection based on summing the currents entering and leaving a node such as a switchboard, bus-bar or a generator winding.

– Current transformers are used to monitor the current entering and leaving the zone to be protected. Provided there is no fault path within the zone the currents will sum to zero.

– If a fault occurs this will no longer be true and a difference signal will be generated operating the over current trip on the circuit breaker.

– Differential protection can be used to create zones around individual bus sections in a multi-split redundancy concept connected as a ring. With this arrangement only the faulty bus section is tripped and all other bus sections remain connected. This has advantages if some of the bus sections do not have a generator connected.

– Differential protection schemes can have problems with high levels of through-fault current. That is current passing through a healthy zone on its way to a fault in some other zone. There have been problems with healthy zones tripping causing failure effects exceeding WCFDI. It is for this reason that some designers favor arc protection for this application.

The effectiveness of differential protection for bus bar applications is difficult to establish conclusively without conducting short circuit testing.

Differential protection is almost universally applied for the protection of generator windings on machines above about 1.5MVA.

6.2.3 (MTS9.7.5) Directional over current protection: Directional over current protection is sometimes applied for bus-bar protection. It is less expensive than differential, due to the reduced number of current transformers required to define a protection zone. Directional over current generally cannot be used with ring configurations as it depends on blocking the upstream circuit breaker from tripping.

6.2.4 (MTS9.7.6) Earth fault protection: The size of the power distribution system and the maximum prospective earth fault current influences the type of earth fault protection specified for marine system.

– Low voltage marine power systems are often designed as un-intentionally earthed systems where the power system has no direct connection or reference to earth (vessel's hull). On these systems, earth faults are typically indicated by earth fault lamps or meters connected from each line to earth.

– Intentional earth impedance should be considered in the case of high voltage systems. High resistance earthing (Grounding) of various types is generally employed.

– All power systems are referenced to earth by way of the distributed capacitance of cables and windings. A significant earth fault current can flow even in unintentionally earthed HV systems.

– The intentional earth impedance adds to the system charging current when an earth fault occurs and should be sized to provide an earth fault current three times that which would flow as a result of the capacitive charging current. This provides well defined current paths for protection purposes.

– Earth fault protection for the main power system is sometimes based solely on time grading. The relay in the earthing (Grounding) resistor or earthing (Grounding) transformers for each bus will detect an earth fault at any point in the plant not isolated by a transformer.

– Earth fault protection in the feeders is used to isolate a fault in a consumer. If the earth fault persists after the tripping time of the feeder the fault is assumed to be in the generators or on the bus-bars itself. At this point the protection driven from the neutral earthing (Grounding) transformers will trip the main bus-ties to limit the earth fault to one bus or the other. Whichever neutral earthing (Grounding) transformer continues to detect an earth fault will then trip all generators connected to that bus. Losing a whole bus due to an earth fault in one generator is unnecessarily severe. Design should consider adding restricted earth fault protection to the generators.

6.2.5 (MTS9.7.7) Over under voltage: This protection element is often a class requirement. It assists in preventing equipment damage but does not contribute to redundancy concept directly.

6.2.6–There should be other protective functions to prevent the power plant reaching the point at which this protection operates. Over/under voltage protection is not selective and blackout is the likely outcome.

To prevent blackout in common power systems (closed bus), design should provide other protective functions which detect the onset of the voltage excursion and divide the common power system into independent power systems or isolate the sources of the fault before healthy generators are tripped (for example a faulty generator).

Operating the power system as two or more independent power system (bus-ties open) provides protection against this fault.

6.2.6 (MTS9.7.8) Over under frequency: Under frequency can be caused by system overload and there must be means of preventing the power plant reaching this condition. Such functions are normally found in the DP control system, power management system, thruster drives and other large drives. Over frequency can be caused by a governor failing to the full fuel condition. This will cause a severe load sharing imbalance which can drive up the bus frequency to the point where several healthy generators trip on over frequency or reverse power. The failure scenarios are similar to those for over and under voltage as described above.

6.2.7 (MTS9.7.9) Reverse power: This protective function is applied to prevent a diesel generator that has lost power from becoming an unacceptable burden on other generators operating in parallel. If a generator with a fuel supply problem sheds the load it is carrying it will be motored by other generators. The power required to motor the faulty generator adds to the load on the healthy generators.

– Although the reverse power trip is a useful function, it makes healthy generators vulnerable to being forced to trip on reverse power if a faulty generator takes all the load. In this failure scenario the healthy generators all trip on reverse power and the faulty set trips on some other protective function leading to blackout.

– Vessels operating their power plant as a common power system should have a means to detect the onset of a generator fault which could have this effect and either subdivide the power plant into independent power systems or trip the generator that is creating the problem.

– Operating the power plant as two or more independent power systems (bus-ties open) provides protection against this type of failure.

6.2.8 (MTS9.7.10) Field failure: This protective function is designed to prevent a generator with field failure (under excitation) becoming an unacceptable reactive power drain on other generators. However, a generator may also fail due to over excitation. If this happens it may push the operating point of healthy generators into the tripping zone of their field failure protection leading to cascade failure and blackout.

– Vessels operating their power plant as a common power system should have a means to detect the onset of a generator fault which could have this effect and either subdivide the power plant into independent power systems or trip the generator that is creating the problem.

– Operating the power plant as two or more independent power systems (bus-ties open) provides protection against this type of failure

6.2.9 (MTS9.7.11) Negative phase sequence protection: Three phase synchronous generators can only tolerate a limited degree of imbalance in their line currents. Large single phase loads, faulty motors or broken conductors may cause a large imbalance which sets up a backwards rotating field in the generator causing overheating.

Negative Phase Sequence protection is used to trip any generator which has a line current imbalance larger than a defined percentage of the full load current. This protection function is not selective and there is a possibility that all online generators may trip in response to a large negative sequence fault.

Vessels operating their power plant as a common power system should have a means to either subdivide the power plant into independent power systems or trip the circuit (feeder, bus section or generator) that is creating the problem.

Operating the power plant as two or more independent power systems (bus-ties open) provides protection against this type of failure.

6.3 POWER SYSTEM: PROCEDURE FOR PROTECTIVE DEVICE COORDINATION

Correct overcurrent relay application requires knowledge of the fault current that can flow in each part of the network. Since large-scale tests are normally impracticable, system analysis must be used.

– one-line diagram of the power system involved, showing the types and ratings of the protection devices

– impedances of all power transformers, rotating machine, and feeder circuits

– maximum and minimum values of short-circuit currents that are expected to flow through each protection device

– maximum load current through protection devices

– starting current requirements of motors and the starting and locked rotor/stalling times of induction motors

– transformer inrush, thermal withstand, and damage characteristics

– decrement curves showing the rate of decay of the fault current supplied by the generators

– relay settings, which are first determined to give the shortest operating times at maximum fault levels

6.4 FAULT CURRENT CALCULATION GUIDELINES (PER USCG REQUIREMENTS)

§111.52–3 Systems below 1500 Kilowatts

The following short-circuit assumptions must be made for a system with an aggregate generating capacity below 1500 kilowatts, unless detailed computations in accordance with §111.52–5 are submitted:

(a) *The maximum short-circuit current of a direct current system must be assumed to be 10 times the aggregate normal rated generator currents plus six times the aggregate normal rated currents of all motors that may be in operation.*

(b) *The maximum asymmetrical short-circuit current for an alternating current system must be assumed to be 10 times the aggregate normal rated generator currents plus four times the aggregate normal rated currents of all motors that may be in operation.*

(c) *The average asymmetrical short circuit current for an alternating-current system must be assumed to be 8 1/2 times the aggregate normal rated generator currents plus 3 1/2 times the aggregate normal rated currents of all motors that may be in operation.*

6.5 OVERALL PROTECTION SYNOPSIS

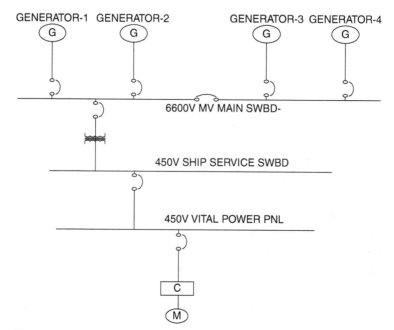

Figure 6.1   Simplified EOL Version-1 Showing Details of Table 6.1

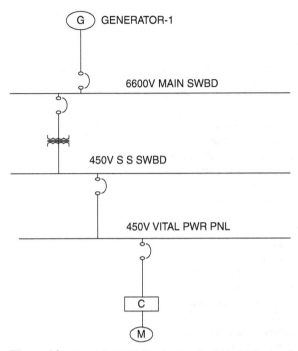

Figure 6.2   Simplified EOL Showing Details Table 6.1 Version-2.

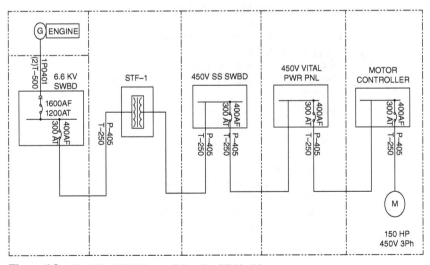

Figure 6.3   Simplified EOL Version-3 Details of Table 6.1

Table 6.2 Explanation for Figure 6.1

Design step#	Item#	Service Description	Remarks
1	17	Motor: Typical ship service load. Each ship service load has its specific category with specific requirements such as emergency fire pump, steering gear pump, air compressor, engine room ventilation, etc. There are specific requirements for grouping these loads so that operational and other requirements can be met.	Design verification is mandatory at this stage
2	16	Feeder cable from the motor to the controller.	
3	15	Motor controller: Provides overload protection by overload relay, manual disconnect, manual start-stop, status indication, and remote control features.	
4	14	Feeder from the controller to the service panel such as power panel, load center, switchboard, etc.	
5	13	Motor protection at the service panel, usually with circuit breaker.	
6	12	Feeder from vital power panel to vital load center.	
7	11	Disconnect means with short circuit protection at the vita load center.	
8	10	RMS symmetric current.	
9	9	Feeder from the vital load center to the ship service switchboard.	
10	8	Disconnect means with short-circuit protection at the ship service switchboard.	
11	7	Feeder from the ship service switchboard to the ship service transformer.	
12	6	Ship service transformer.	
13	5	Transformer feeder to the MV switchboard.	
14	4	Protective device at the MV switchboard for the ship service transformer.	
15	3	Protective device for the generator at the MV switchboard.	
16	2	Generator feeder.	
17	1	Generator.	

6.6 ANSI ELECTRICAL DEVICE NUMBERING (FOR DEVICE NUMBER DETAILS REFER TO ANSI C.37.2)

Table 6.3   List of ANSI Electrical Devices (Partial)

ANSI Device	ANSI Device Description	Remarks
24	Over-Excitation Protection	Over-Excitation Protection
25	Synchronization Relay	
27	Under Voltage Protection Relay	
32	Directional Power Relay	Directional Power Relay 32P, P→, - active power 32Q, Q→ - reactive power
40U & 40O	Field (Over/Under Excitation) Relay	
43	Manual Transfer or Selector Device	
44	Unit Sequence Starting Relay	
46	Reverse-Phase or Phase-Balance Current Relay	Negative-Phase Sequence Current Relay
47	Phase-Sequence or Phase-Balance Voltage Protection Relay	
48	Incomplete Sequence Relay	
49	Machine or Transformer, Thermal Relay	
49F	Thermal Protection for Cable	
49M, 49G, 49T	Three-Phase Thermal Protection for Machines 49M for motor, 49G for generators, 49T for transformer	
50	Instantaneous Overcurrent Relay	Instantaneous Overcurrent Relay, or rate-of-rise relay that functions instantaneously on an excessive value of current, or on an excessive rate of current rise, thus indicating a fault in the apparatus of the circuit being protected.
51	AC Inverse Time Overcurrent Relay	AC Time Overcurrent Relay, which is a relay with either a definite or inverse time characteristic that functions when the current in the AC circuit exceeds a predetermined value.
50N/51N	Non-Directional Ground Fault	Earth Fault Protection, based on measured or calculated residual current values by 3-phase current sensors
50G/51G		Earth Fault Protection, based on measured or calculated residual current values measured directly by a specific sensor.

(cont.)

Table 6.3   (*Cont.*)

ANSI Device	ANSI Device Description	Remarks
52	Power Circuit Breaker	
53	Exciter or DC Generator R	
55	Power Factor Relay	
56	Field Application Relay	
57	Short-Circuiting or Grounding Device	
59, 59N	Overvoltage Protection Relay	
60	Voltage or Current Balance Relay	
60LOP	Loss of Potential Logic	
64	Ground Detection Relay	
67	Directional Overcurrent 67N, I0 > → - directional earth-fault	
68	Transformer/Motor Inrush Current	
78	Phase-Angle Measuring Relay	
79	AC Reclosing Relay-Auto Reclosure	
81, 81R	Frequency Relay Rate of Frequency Protection	81N, f < - under frequency 81O, f > - over frequency
86	Lockout Relay	
87G, 87M, 87T	Differential Protection	Differential Protection 87G, $\Delta I >$ - generator 87M, $\Delta I >$ - motor 87T, $\Delta I >$ - transformer

ANSI 50N/51N or 50G/51G – Earth fault

Earth fault protection based on measured or calculated residual current values:

- ANSI 50N/51N: residual current calculated or measured by 3-phase current sensors
- ANSI 50G/51G: residual current measured directly by a specific sensor

6.7 FAULT CURRENT CALCULATIONS (PER USCG REQUIREMENTS CFR 111-52-3(B) & (C))

Example-1: The following fault current calculation is performed for a maximum of two generators (1000 kW each) operating in parallel and with an electric plant motor load that is assumed to be 1200 kW.

The short-circuit current is calculated as follows:

(a) **Calculate generator full load current: (Two generators rated 1000 kW each)**

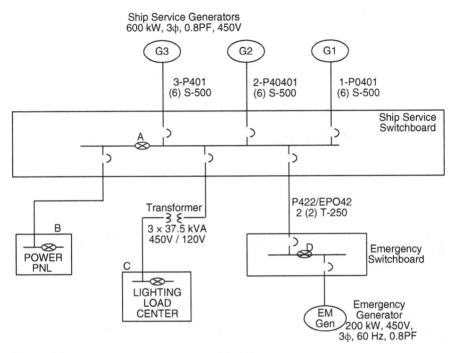

Figure 6.4 Simplified EOL for Short-Circuit Calculation.

Each generator full load current $= (1000 \times 1000) \ / \ (1.732 \times 450 \times 0.8) = 1604$ AMP.

(b) Total Motor contribution in amps: (1200 kW)
Motor running current $= (1200 \times 1000) \ / \ (1.732 \times 450 \times 0.8 \times 0.9) = 2138$ AMP.

(This assumes that the motors are approximately 90% efficient and are designed for a 0.8 power factor).

6.7.1 Maximum Asymmetrical Fault Current

The maximum asymmetrical short-circuit current (2 generators in parallel) $= (2 \times 10 \times 1604) + 4 \times 2138 = 32{,}080 + 8552 = 40{,}632$ AMP.

6.7.2 Average Asymmetrical Fault Current

The average asymmetrical current is $(8.5 \times (2 \times 1604) + 3.5 \times 2138) = 27268 + 7483 = 34{,}751$ AMP.

6.7.3 450 V Switchboard Rating

The 450 V, AC, 60 Hz, 3-phase, ship service switchboard rating shall be to withstand the maximum asymmetrical short-circuit current of the system. The calculated maximum asymmetrical short-circuit is 40,632 AMP. Therefore, the 450 V ship service switchboard rating shall be at least 42 KAIC for this installation.

6.7.4 450 V Switchboard Circuit Breaker Rating

The preceding calculation also provides the switchboard circuit breaker rating requirements. All 450 V ship service switchboard breakers shall be rated for minimum of 42 KAIC.

6.7.5 Fault Current Calculation for the 120 Voltage System is as follows

For a 400 kVA step down transformer 450 V/120 V:

For 120 V low-voltage system circuit breaker ratings (connected through step down transformers), the transformer rating and its low-voltage rated current need to be calculated.

6.7.6 RMS Symmetric Current

The *RMS* symmetrical current can be estimated by multiplying the low-voltage current by 20. The asymmetrical current values may then be obtained by applying the same factors as above.

For example the RMS symmetrical current for a 3-phase 120 volt switchboard connected through a 250 kVA transformer would be:

$$= 20 \times 250 \times 1000/(1.732 \times 120)$$
$$= 24,056 \text{ AMP.}$$

6.7.7 Fault Current Calculation Summary

Table 6.4 Calculated Fault Current Summary (Two 1000 kW Generators Running Parallel)

Calculated fault current summary (Two 1000 kW generators in parallel)		
Item		
1	Two 600 kW generators in parallel	Maximum asymmetrical current: 40,632 AMP	Average Asymmetrical current: 34,751 AMP
2	450 V Switchboard Rating	42 KAIC	
3	120 V Panel board Rating	32 KAIC	

Summary:

a. The initial symmetrical, three-phase short circuit current value at the ship service switchboard bus is 19,520 AMP and the line-to-line fault is 16,950 AMP.

b. The asymmetric, three phase short circuit current value at the ship service switchboard bus is 27,902 AMP and the line-to-line fault is 24,177 AMP.

c. Based on Fault Current Analysis by both USCG regulations and the SKM software method; it is recommended that the ship service bus bar and bracing be suitable to withstand 30,000 AMP.

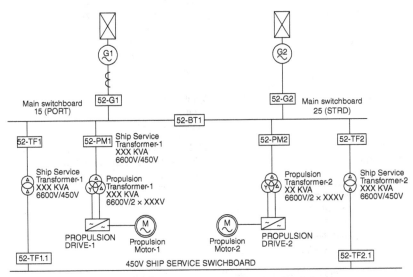

Figure 6.5   Typical MV Generator Protection Scheme for One Bustie Breaker.

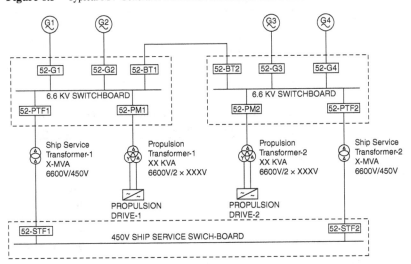

Figure 6.6   Typical EOL for MV Generator Protection System: Spilt Bus with Two Bustie Breakers.

Table 6.5 Typical EOL for MV Generator Protection System: Spilt Bus with Two Bustie Breakers

Generator rating: 5 MW, 0.8 PF, 6.6 kV, 60 Hz, 540 AMP
Ship service transformer rating:

(A) Circuit Breaker Protection ANSI Device Details

Device#	Device Description	Function	Remarks
52-G1 & 52-G2	Circuit breaker	Generator 1 and generator 2 circuit breakers	630 AF
52-BT1 & 52-BT2	Circuit breaker	Bustie-1 and bustie-2 circuit breakers	1250 AF
52-PTF1 & 52-PTF2	Circuit breaker	Transformer primary circuit breakers 1 & 2	630 AF
52-STF1.1 & 52-STF2.1	Circuit breaker	Transformer secondary circuit breakers 1 & 2	4800 AF
52-PM1 & 52-PM2	Circuit breaker	Propulsion drive transformer primary circuit breakers 1 & 2	

(B) Generator Protection—Lockout Protection VIA Lockout Relay 86. Lockout is to Ensure that the Fault is Cleared and the Lockout Relay Manually Reset at the HV Switchboard Before the Circuit Breaker Can Be Reclosed

51	Phase Overcurrent Relay
87G	Generator Differential Protection
40	Loss of Field
67N	Directional Earth Fault Relay
51N	Grounding Protection

(C) Generator Protection for Direct Tripping of the Generator Circuit Breaker (no Lockout)

27	Under Voltage
59	Over Voltage
81U/O	Under Frequency and Over Frequency
46	Negative Phase Sequence
32R	Reverse Power
24	Over Excitation

(D) Bus-Tie Interconnection Protection with Lockout Features with 86 Lockout Relay (Manual Resettable)

67	Directional Phase Overcurrent	
50	Instantaneous Overcurrent Relay	
67N		
50N		For Cable Interconnection (Generator Feeder) Only
50N		For Bus-Tie Only

Table 6.5 (*Cont.*)

(E) For Cable Interconnection /Bus-Tie Circuit Breaker Protection			
Device#	Device Description	Function	Remarks
27	Under Voltage		For Circuit Breaker Trip
59	Over Voltage		For Circuit Breaker Trip
81U/O	Under/Over Frequency		For Circuit Breaker Trip

(F) For Ship Service Transformer (6.6 KV/450 V) Primary Circuit Breaker Protection with Lockout Feature (Relay 86)			
51	Phase Overcurrent—IDMT		
50	Definite Time High Set Overcurrent		Inrush Blocking
50N	Definite Time Ground protection		

6.8 DETAILS FOR FIGURE 6.3 TYPICAL EOL FOR MV GENERATOR PROTECTION SYSTEM: SPLIT BUS WITH TWO BUSTIE BREAKERS

Table 6.6 Typical EOL For MV Generator Protection System: Split Bus with Two Bustie Breakers

Generator Protection with Discrete or Multifunction Relay (Figure 6.7)			
Device#	Device Description	Function	Remarks
52-G1 & 52-G2	Circuit breaker	Generator 1 and generator 2 circuit breakers	
52-BT1 & 52-BT2	Circuit breaker	Bustie-1 and bustie-2 circuit breakers	
52-PTF1 & 52-PTF2	Circuit breaker	Transformer primary circuit breakers 1 & 2	
52-STF1.1 & 52-STF2.1	Circuit breaker	Transformer secondary circuit breakers 1 & 2	
52-PM1 & 52-PM2	Circuit breaker	Propulsion drive transformer primary circuit breakers 1 & 2	

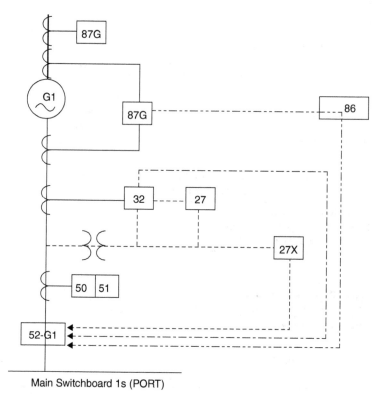

Main Switchboard 1s (PORT)

Figure 6.7 Typical MV Generator Protection Schematic.

Table 6.7 For Figure 6.4 Typical EOL for MV Generator Protection System: Split Bus with Two Bustie Breakers

Generator Protection with Discrete or Multifunction Relay (Figure 6.4)			
ANSI Device#	Device Description	Function	Remarks
27	Under Voltage		
32	Directional Power Relay		
50	Instantaneous Overcurrent Relay		
51	Phase Overcurrent—IDMT		
86			
87G	Generator Differential Protection		

6.9 DETAILS FOR FIGURE 6-4: TYPICAL EOL FOR MV GENERATOR PROTECTION SYSTEM: SPLIT BUS WITH TWO BUSTIE BREAKERS

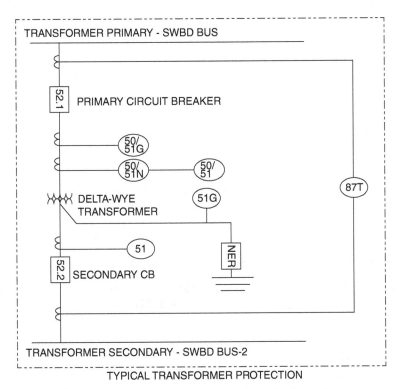

TRANSFORMER PRIMARY - SWBD BUS

52.1 PRIMARY CIRCUIT BREAKER

50/51G

50/51N   50/51

DELTA-WYE TRANSFORMER   51G   87T

51   NER

52.2 SECONDARY CB

TRANSFORMER SECONDARY - SWBD BUS-2

TYPICAL TRANSFORMER PROTECTION

Figure 6.8 Typical MV Transformer (Delta-WYE) Protection Schemetic.

Table 6.8 Typical MV Transformer (Delta-WYE with Neutral) Protection Schematic (for Figure 6.5).

Transformer Protection with Discrete or Multifunction Relay (Figure 6.5)			
ANSI Device#	Device Description	Function	Remarks
27	Under Voltage		
32	Directional Power Relay		
50	Instantaneous Overcurrent Relay		
51	Phase Overcurrent—IDMT		
51N	Non-Directional Ground Fault	Earth fault protection based on measured or calculated residual current values by 3-phase current sensors	

Table 6.8   (*Cont.*)

	Transformer Protection with Discrete or Multifunction Relay (Figure 6.5)		
ANSI Device#	Device Description	Function	Remarks
51G		Earth fault protection based on measured or calculated residual current values measured directly by a specific sensor	
87T	Differential Protection:Transformer		
50N/51N	Non-Directional Ground Fault	Earth fault protection based on measured or calculated residual current values by 3-phase current sensors	
50G/51G		Earth fault protection based on measured or calculated residual current values measured directly by a specific sensor	

6.10 DETAILS FOR FIGURE 6.5 TYPICAL FOR TRANSFORMER PROTECTION SCHEMATIC

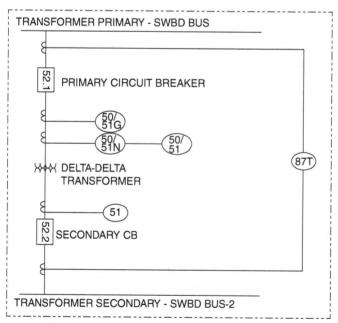

Figure 6.9   Typical MV Transformer (Delta-WYE) Protection Schematic.

Table 6.9   For Figure 6.9: Typical EOL for MV Generator Protection Syastem: Split Bus with Two Bustie Breakers

Transformer Protection with Discrete or Multifunction Relay (Figure 6.6)			
ANSI Device#	Device Description	Function	Remarks
27	Under Voltage		
32	Directional Power Relay		
50	Instantaneous Overcurrent Relay		
51	Phase Overcurrent—IDMT		
51N		Earth fault protection based on measured or calculated residual current values by 3-phase current sensors	
51G		Earth fault protection based on measured or calculated residual current values measured directly by a specific sensor	
87T	Differential Protection: Transformer		

6.11 DETAILS FOR FIGURE 6.10: TYPICAL EOL FOR MV VFD TRANSFORMER PROTECTION SCHEMATIC

6600V MAIN SWITCHBOARD-FEEDER FOR PROPULSION MOTOR (PM) DRIVE

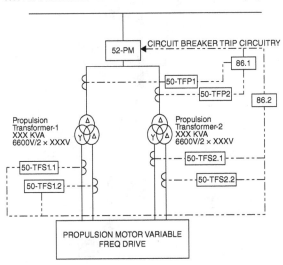

50-TFP1, 50-TFP2 - TRANSFORMER PRIMARY OVER CURRENT RELAY
50-TFS1.1, 50-TFS1.2, 50-TFS2.1, 50TFS2.2-ALL FOR TRANSFORMER
SECONDARY PROTECTION

86.1 AND 86.2 LOCKOUT RELAY

Figure 6.10   Typical MV Transformer VFD FEED Protection Schemetic.

Table 6.10 For Figure 6.10: Typical EOL for MV Generator Protection Syastem: Split Bus with Two Bustie Breakers

Transformer Protection with Discrete or Multifunction Relay (Figure 6.6)

ANSI Device#	Device Description	Function	Remarks
27	Under Voltage		
32	Directional Power Relay		
50	Instantaneous Overcurrent Relay		
51	Phase Overcurrent—IDMT		
51N		Earth fault protection based on measured or calculated residual current values by 3 phase current sensors	
51G		Earth fault protection based on measured or calculated residual current values measured directly by a specific sensor	
87T	Differential Protection-Transformer		

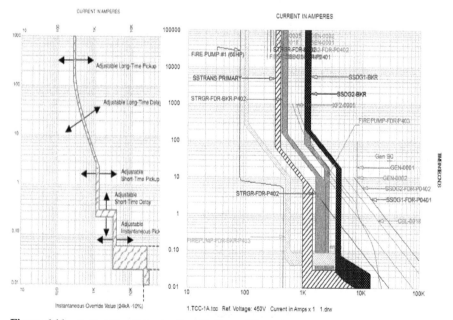

Figure 6.11 Typical Circuit Breaker Coordination and Protection.

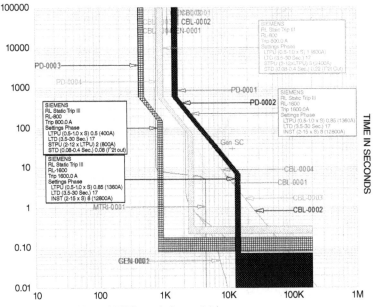

Figure 6.12 Typical Shipboard Motor Protection-1.

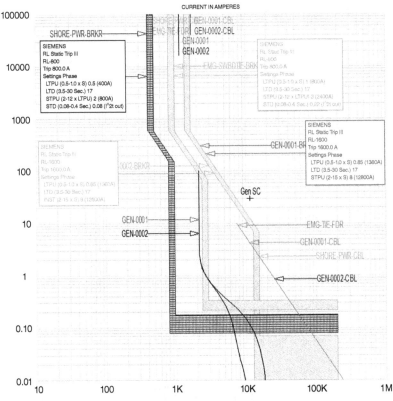

Figure 6.13 Typical Shipboard Motor Protection-1.

CURRENT IN AMPERES

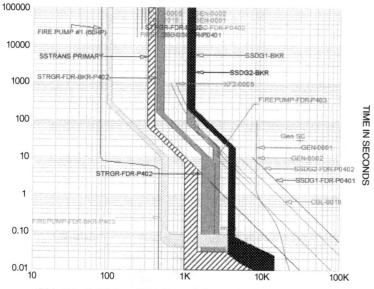

1. TCC-1A.tcc Ref. Voltage: 450V Current in Amps x 1 1.drw

Figure 6.14 Typical Shipboard Motor Protection-1.

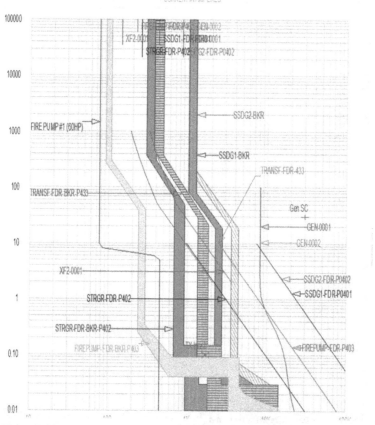

Figure 6.15 Typical Shipboard Motor Protection-1.

- Calculate setting of IDMT Overcurrent Relay for following Feeder and CT Detail
- Feeder Detail: Feeder Load Current 384 AMP, Feeder Fault current Min11KA and Max 22KA.
- CT Detail: CT installed on feeder is 600/1 AMP. Relay Error 7.5%, CT Error 10.0%, CT over shoot 0.05 Sec, CT interrupting Time is 0.17 Sec and Safety is 0.33 Sec.
- IDMT Relay Detail:
- IDMT Relay Low Current setting: Over Load Current setting is 125%, Plug setting of Relay is 0.8 AMP and Time Delay (TMS) is 0.125 Sec, Relay Curve is selected as Normal Inverse Type.
- IDMT Relay High Current setting: Plug setting of Relay is 2.5 AMP and Time Delay (TMS) is 0.100 Sec, Relay Curve is selected as Normal Inverse Type

Calculation of Over Current Relay Setting:

6.11.1 Low Overcurrent Setting: (I >)

- Overload Current (In) = Feeder Load Current × Relay setting = $384 \times 125\% = 480$ AMP (125% REQUIREMENT ???)
- 1.1 Required Over Load Relay Plug Setting = Over Load Current (In) / CT Primary Current
- Required Over Load Relay Plug Setting = $480 / 600 = 0.8$
- 1.2 Pickup Setting of Overcurrent Relay (PMS) (I >) = CT Secondary Current X Relay Plug Setting
- Pickup Setting of Overcurrent Relay (PMS) (I >) = $1 \times 0.8 = 0.8$ AMP
- 1.3 Plug Setting Multiplier (PSM) = Min. Feeder Fault Current / (PMS × (CT Pri. Current / CT Sec. Current))
- Plug Setting Multiplier (PSM) = $11000 / (0.8 \times (600 / 1)) = 22.92$
- Operation Time of Relay as per its Curve:
- Operating Time of Relay for Very Inverse Curve (t) = $13.5 / ((PSM)-1)$.
- Operating Time of Relay for Extreme Inverse Curve (t) = $80/ ((PSM)2-1)$.
- Operating Time of Relay for Long Time Inverse Curve (t) = $120 / ((PSM)-1)$.
- Operating Time of Relay for Normal Inverse Curve (t) = $0.14 / ((PSM) 0.02-1)$.
- Operating Time of Relay for Normal Inverse Curve (t) = $0.14 / ((22.92) 0.02-1) = 2.17$ AMP
- Here Time Delay of Relay (TMS) is 0.125 Sec so
- 1.4 Actual operating Time of Relay (t >) = Operating Time of Relay × TMS (TIME DELAY) = $2.17 \times 0.125 = 0.271$ Sec

- 1.5 Grading Time of Relay $= [((2 \times \text{Relay Error}) + \text{CT Error}) \times \text{TMS}] + \text{Over shoot} + \text{CB Interrupting Time} + \text{Safety}$
- Total Grading Time of Relay $= [((2 \times 7.5) + 10) \times 0.125] + 0.05 + 0.17 + 0.33 = 0.58$ Sec
- 1.6 Operating Time of Previous Upstream Relay = Actual operating Time of Relay + Total Grading Time Operating Time of Previous Upstream Relay $= 0.271 + 0.58 = 0.85$ Sec

6.11.2 High Overcurrent Setting: (I >>)

2.1 Pickup Setting of Overcurrent Relay (PMS) (I >>) = CT Secondary Current X Relay Plug Setting

Pickup Setting of Overcurrent Relay (PMS) (I >) $= 1 \times 2.5 = 2.5$ AMP

2.2 Plug Setting Multiplier (PSM) = Min. Feeder Fault Current / (PMS $\times$ (CT Pri. Current / CT Sec. Current))

Plug Setting Multiplier (PSM) $= 11000 / (2.5 \times (600 / 1)) = 7.33$

Operation Time of Relay as per it's Curve

Operating Time of Relay for Very Inverse Curve (t) $= 13.5 / ((\text{PSM}) - 1)$.

Operating Time of Relay for Extreme Inverse Curve (t) $= 80 / ((\text{PSM})2 - 1)$.

Operating Time of Relay for Long Time Inverse Curve (t) $= 120 / ((\text{PSM}) - 1)$.

Operating Time of Relay for Normal Inverse Curve (t) $= 0.14 / ((\text{PSM}) 0.02 - 1)$.

2.3 Operating Time of Relay for Normal Inverse Curve (t) $= 0.14 / ((7.33) 0.02 - 1) = 3.44$ AMP

Here Time Delay of Relay (TMS) is 0.100 Sec so

2.4 Actual Operating Time of Relay (t >) = Operating Time of Relay $\times$ TMS $= 3.44 \times 0.100 = 0.34$ Sec

2.5 Grading Time of Relay $= [((2 \times \text{Relay Error}) + \text{CT Error}) \times \text{TMS}] + \text{Over shoot} + \text{CB Interrupting Time} + \text{Safety}$

Total Grading Time of Relay $= [((2 \times 7.5) + 10) \times 0.100] + 0.05 + 0.17 + 0.33 = 0.58$ Sec

2.6 Operating Time of Previous Upstream Relay = Actual Operating Time of Relay + Total Grading Time.

Operating Time of Previous Upstream Relay $= 0.34 + 0.58 = 0.85$ Sec

6.11.3 Conclusion of Calculation

Pickup setting of overcurrent relay (PMS) (I >) should be satisfied following two conditions.

(1) Pickup setting of overcurrent relay (PMS)(I >) > = Over Load Current (In) / CT Primary Current

(2) TMS $\Leftarrow$ Minimum Fault Current / CT Primary Current

For condition (1) 0.8 > = (480/600) = 0.8 > = 0.8, Which found **OK**

For condition (2) 0.125 ⇐ 11000/600 = 0.125 ⇐ 18.33, Which found **OK**

Here conditions (1) and (2) are satisfied so

Pickup setting of overcurrent relay = OK

Low overcurrent relay setting: (I >) = 0.8A × In AMP

Actual operating time of relay (t >) = 0.271 Sec

High overcurrent relay setting: (I >>) = 2.5A × In AMP

Actual operating time of relay (t >>) = 0.34 Sec

Zone 1–Input transformer

Differential, overcurrent protection.

Based upon the transformer connection ground over current and neutral voltage detection.

1. Short Circuit/Overcurrent – Fuse, CB or OC relay 2.
2. Overload – overcurrent protection with time delay
3. Voltage Unbalance – loss of input phase
4. Ground Fault Overcurrent

A device 51 (AC Inverse Time Overcurrent Relay) element that operates on the fundamental frequency (i.e. not rms) may be set with a lower pickup, as it will not respond to the harmonic components of the load current. An instantaneous ground element (50N) and differential (87) can be applied. The feeder 51 can provide conventional time delayed protection.

For ASD drives, if employ capacitors for power factor correction, resonances can occur.

Zone 2–Power Electronics (Drives Equipment)

Over- or undervoltages and voltage unbalances.

Some drives may also include DC link reactor overvoltage protection.

Overcurrent protection is provided for the converter electronics and interconnected bus or wiring.

Short-circuit protection is provided by fuses ahead of the thyristors.
   – DC Overvoltage
   – DC Undervoltage - loss of control power
   – Over Temperature – this includes the rectifier and inverter heat sinks as well as the enclosure temperature

Zone 3–Motor

Motors should be provided with the same protection as constant speed motors of the same size.

Motor protection should also include:

Over-frequency and overvoltage or

Over-excitation protection.

This protection is typically provided by the drive control system but could be provided by discreet or multifunction relays.
  – Ground Fault, Motor Overcurrent, Motor Overload, Motor Stall, Motor Overspeed, Current Unbalance
  – Underload – may indicate a process malfunction and will protect the machinery and the process in this fault condition
  – External Fault – an external relay input

Some of the additional protection functions are already included in the ASD drives i.e., off nominal frequency operation.

In particular, attention should be given to the low frequency saturation point of current transformers and the low frequency response characteristics of protective relays placed downstream of the ASD. Thermal models are defined at the nominal frequency. Additional analysis is required to select the thermal 4 model at off nominal frequencies.

Changes to Motor Operating Characteristics and Dynamics that can Impact Protection

When motors are applied to ASDs, certain operating characteristics of the motor are modified. The operating frequency impacts how the motor behaves during operation both starting and running as well as during abnormal operation and fault conditions. The areas that will be discussed are pertinent to the protection of the motor and drive system.

Reduced Frequency Operation Effects:

The frequency of the source to the motor dictates the operating speed. At lower speed operation, the motor is not cooled as efficiently as it is at rated speed. Therefore this must be taken into consideration with regard to motor thermal overload protection. Actual motor full load current (FLA) is a function of the frequency, as lower FLA is drawn at lower frequency. The actual FLA must be used in overload protection.

Harmonics in the motor current will cause additional heating in the motors and other connected elements. This additional heating needs to be considered when sizing and protecting the equipment. At near rated load, a typical value to accommodate the additional heating can be up to 15 percent more above the fundamental heating effects.

Voltage and Dielectric Stresses:

– Consideration should be given for overvoltage protection due to the potential for this condition and the fact that many drives operate at fairly high semiconductor switching frequencies. The concern is especially important with drives applied on long cable runs that can have high voltages developed due to cable capacitance. Sustained overvoltage conditions can be detected by overvoltage protection functions but the ringing effect must be mitigated by other voltage control methods.

ASD to include: Converter, Inverter, Current Source Inverter (CSI) & Voltage Source Inverter (VSI), Current–Source Converter & Voltage–Source Converter, Pulse Width Modulation (PWM)

Zone 1–Input Transformer

Differential, and overcurrent protection.
    Based upon the transformer connection ground Overcurrent and neutral voltage detection.

1. Short Circuit/Overcurrent – Fuse, CB or OC relay
2. Overload – Overcurrent protection with time delay
3. Voltage Unbalance – Loss of input phase
4. Ground Fault Overcurrent

An ANSI device 51 (Time Overcurrent) element that operates on the fundamental frequency (i.e., not rms) may be set with a lower pickup, as it will not respond to the harmonic components of the load current. An instantaneous ground element (50N) and differential (87) can be applied. The feeder 51 can provide conventional time delayed protection. For ASD drives, if employ capacitors for power factor correction, resonances can occur, which must be taken under consideration.

Zone 2–Power Electronics (Drives Equipment)

Over- or undervoltages and voltage unbalances will be addressed. Some drives may also include DC link reactor overvoltage protection. Overcurrent protection is provided for the converter electronics and interconnected bus or wiring. Short-circuit protection is provided by fuses ahead of the thyristors.

1. DC Overvoltage
2. DC Undervoltage – loss of control power
3. Over Temperature – this includes the rectifier and inverter heat sinks as well as the enclosure temperature

Zone 3–Motor

Motors should be provided with the same protection as constant speed motors of the same size. Motor protection should also include overfrequency and overvoltage or overexcitation protection. This protection is typically provided by the drive control system but could be provided by discreet or multifunction relays.

1. Motor Ground Fault
2. Motor Overcurrent
3. Motor Overload I^2t
4. Motor Stall
5. Motor Overspeed
6. Current Unbalance
7. Underload – may indicate a process malfunction and will protect the machinery and the process in this fault condition
8. External Fault – an external relay input

Changes to Motor Operating Characteristics and Dynamics that can Impact Protection

When motors are applied to ASDs, certain operating characteristics of the motor are modified. The operating frequency impacts how the motor behaves during operation both starting and running as well as during abnormal operation and fault conditions.

Reduced Frequency Operation Effects: The frequency of the source to the motor dictates the operating speed. At lower speed operation the motor is not cooled as efficiently as it is at rated speed. Therefore this must be taken into consideration with regard to motor thermal overload protection. The actual motor full load current (FLA) is a function of the frequency, as a lower FLA is drawn at a lower frequency. The actual FLA must be used in overload protection.

Harmonics: Harmonics in the motor current will cause additional heating in the motors and other connected elements. This additional heating needs to be considered when sizing and protecting the equipment. At near rated load, a typical value to accommodate the additional heating can be up to 15 percent increase above the fundamental heating effects.

IEEE Std 242-2001 (Buff Book): Recommended Practice for Protection and Coordination of Industrial and Commercial Power Systems, Section 8.2.5
  If this ground fault is intermittent or allowed to continue, the system could be subjected to possible severe over-voltages to ground, which can be as high as six to eight times phase voltage. Such over-voltages can puncture insulation and result in additional ground faults. These over-voltages are caused by repetitive charging of the system capacitance or by resonance between the system capacitance and the inductance of equipment in the system.

IEEE Std 242-2001 (Buff Book): Recommended Practice for Protection and Coordination of Industrial and Commercial Power Systems 8.2.5

If this ground fault is intermittent or allowed to continue, the system could be subjected to possible severe over-voltages to ground, which can be as high as six to eight times phase voltage. Such over-voltages can puncture insulation and result in additional ground faults. These over-voltages are caused by repetitive charging of the system capacitance or by resonance between the system capacitance and the inductance of equipment in the system.

6.12 POWER SYSTEM DYNAMIC CALCULATIONS

NORSOK Standard Electrical Systems E-CR-001 Rev. 2, January 1996:

A stability analysis of the electrical power system shall be carried out and shall comprise system transient behavior following disturbances during relevant operational modes of the installation.

The simulations shall include:

A. Generator short-circuit with clearance of the fault and generator trip after the set time delay of the protective relays. Based on the analysis, load shedding shall be applied when required.

B. Generator trip. Based on analysis, load shedding shall be applied when required. The analysis shall be carried out for the worst-case conditions with respect to system stability, which shall be determined separately by each project. The analysis shall prove that the system will re-stabilize following the specified disturbances, and that the transient voltage and frequency variations, motor slip, reacceleration and start up times are within acceptable limits.

C. Protective relay coordination and discrimination study.

D. Short-circuited feeders with clearance of the fault after set time delay of the protective relays or blowing time of the fuses.

E. Establish individual system level line to ground capacitance.

F. VFD application with isolation transformer.

G. VFD Active Front End (AFE) application.

H. Power Filter application – EMC.

I. Power Filter application – Low Pass Filter.

J. Power Filter application – High Pass filter.

K. Direct on line starting of the largest motors.

6.13 PROTECTIVE RELAY COORDINATION AND DISCRIMINATION STUDY

A relay coordination study shall be carried out to determine the settings of the protective relays and direct acting circuit breakers.

Series connected overcurrent relays, direct acting circuit breakers and fuses shall be coordinated to achieve correct discrimination during fault conditions. Correct discrimination shall be maintained for the minimum and maximum prospective fault currents, while the thermal effect of the fault current shall not exceed the thermal withstand capability of any circuit component.

The relay coordination study shall be carried out according to the requirements of the IFEA, "Guidelines for the Documentation of Selectivity (Discrimination) in AC Systems."

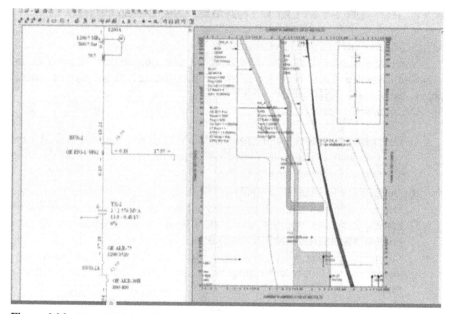

Figure 6.16   Typical Shipboard Motor Protection-1.

FPN-6-1: GENERATOR PROTECTION WITH DISCRETE OR MULTI-FUNCTION RELAY (FIGURE 6.7)

FPN-6-2: For the Generators:

> **FPN-6.2-1:** The generator steady state short circuit current maintained at a minimum 300% full load current $(3 \times 504) = 1512$ amp at a minimum of 1 sec, the generator over-current protection (51) is set to trip the circuit breaker in less than 1 sec to take account of discrimination with down stream circuits.

> **FPN-6.2-2:** The generator and interconnecting generator feeder cable between the generator and its circuit breaker are provided with differential protection (87G). The protection is set to provide the maximum protection for generator internal faults with a low protection setting

(10% differential limit). With this setting the interconnecting cable is also protected for very low fault current with very fast tripping of the circuit breaker to isolate the faulty circuit from other healthy generators operating in parallel.

FPN-6.2-3: The high set phase overcurrent (50) setting is selected to take account of the busbar protection scheme to allow adequate discrimination for the bus bar fault.

FPN-6.2-4: Under voltage, over voltage, under frequency and over frequency settings are selected to provide back-up protection to the protection settings selected on generator feeder cable protection to ensure any down stream sudden abnormal operation will first trip the circuit breaker associated with cable interconnection to avoid black-out.

FPN-6.2-5: Reverse power protection for each generator, when arranged for parallel operation, is set around 10% for around 5 sec.

Chapter 7

Power Quality: Harmonics

7.0 INTRODUCTION

The design and development of shipboard power generation and distribution systems is in accordance with rules and regulations to ensure power source characteristics are suitable for the intended ship service and emergency applications. Power systems are classified in many different categories such as type-I, type-II, emergency, essential, nonessential, etc. The commercial power characteristics started with MIL-STD-1399 with generalized voltage and frequency tolerances as well as total harmonic distortion limits. An all-electric ship power system with variable frequency drive and adjustable speed drive generates electric noise in many different forms, influencing the power quality in the ship's service power distribution system. The shipboard power system harmonic noise in the electrical system requires better understanding of the harmonic generation, propagation, and power system contamination issues so that necessary steps are taken to bring the power system THD level to the acceptable level. The variable frequency drives and adjustable speed drives come with various configurations, such as 6-pulse drive, 12-pulse drive, 18-pulse drive, active front end, pulse width modulation, and many others.

Electrical power system total harmonic distortion (voltage and current) calculation software is commercially available to determine various drive configurations, such as 6-pulse, 12-pulse, and 18-pulse systems. The drive selection is very complex due to fact that system level applications are not often realized during the preliminary and detailed design. Therefore, harmonic noise management to meet the regulatory body requirements is used, such as a low pass power system filter, band pass power system filter, hybrid power system filter, active power system filter, and combinations thereof. Some shipboard power distribution systems require systematic power quality verification so that the requirements for the quality of power are strictly adhered to.

Shipboard electrical power generation and distribution is a completely islanded system. The shipboard power generation and consists of N + 1 generators and one emergency generation. Other marine applications such as offshore floating platforms also use a similar concept. Power generation and distribution system requirements must be in compliance with USCG regulations, ABS rules, IEEE recommendations, and many other applicable national and international standards as appropriate.

Shipboard Power Systems Design and Verification Fundamentals, First Edition. Mohammed M. Islam.
© 2018 the Institute of Electrical and Electronics Engineers, Inc. Published 2018 by John Wiley & Sons, Inc.

Table 7.1 Motor Starter Inrush Current Requirements

Motor Starter Type	Motor Starting Current (% of Full Load Current)	Remarks
Across the line starter	600%–800%	Starting inrush current is the highest
Autotransformer starter	400%–500%	Starting inrush current requirements
Wye/delta starter	200%–275%	Starting inrush current requirements
Solid-state soft starter	200%	Starting inrush current requirements
VFD/ASD	100%	Starting inrush current is the lowest

The use of VFD evolved from various types of starting systems with pros and cons. Table 7.1 provides a list of starting systems with a starting inrush current.

IEEE 519 addresses power system harmonics. IEEE Std 519-2014 is a simplified version of the 1992 version. For shipboard application, IEEE 519 may not provide enough guidelines for voltage and current harmonics calculations and mitigation.

The American Bureau of Shipping's "Guidance Notes on Control of Harmonics in Electrical Power Systems," Pub. 150 EI Harmonics, provides an excellent tutorial for shipboard and offshore platform power generation and distribution-related voltage and current harmonics calculations in different configurations of harmonic generating loads. The document also outlines details of the harmonic mitigating means to comply with the harmonic limitation requirements such as the application of reactors, passive filters, wideband filters, active filters, and hybrid filters at 6-pulse, 12-pulse, and 18-pulse configurations.

The VFD or ASD shipboard application is a very complex design and development process. Harmonic generation as well as mitigation must be phased in at the preliminary design phase by the subject matter experts having jurisdiction in this field. The shipbuilder, the designer, the VFD supplier, and harmonic mitigating experts must work together to establish a proper harmonic management system.

Table 7.2 Typical Generator Reactance Values

Reactance	Symbol	Range	Effective Time	Remarks
Sub-transient reactance	X''_d	0.09–0.21	0 to 6 cycles	Determines maximum instantaneous current
Transient reactance	X'_d	0.13–0.20	6 cycles to 5 sec	Determines short time delay
Synchronous reactance	X_d	1.7–3.3	After 5 sec	Determines steady state current

The following VFD applications are analyzed for design considerations:

(1) 6-pulse drive

(2) 6-pulse drive with reactors

(3) 6-pulse drive with passive filters

Table 7.3   Typical Transformer Impedance Values

Transformer Short-Circuit Impedance at Rated Current (Taken from IEC)		
Rated power in kVA	Minimum short-circuit impedance in %	Remarks
Up to 630 kVA	4.0	
631 to 1250	5.0	
1251 to 2500	6.0	
6301 to 25,000	7.0	

Table 7.4   Typical Voltage Harmonic Orders for VFD(S) and ASD(S)

Drives	Voltage Harmonic Order									Remarks
	5th	7th	11th	13th	17th	19th	23rd	25th	Up to 49th	
6-Pulse	0.175	0.11	0.045	0.029	0.015	0.009	0.008	0.007	$\geq 27\%$	Manage 5th and 7th
12-Pulse	0.026	0.016	0.073	0.057	0.002	0.001	0.02	0.016	$\geq 11\%$	Manage 11th and 13th
18-Pulse	0.021	0.011	0.007	0.005	0.045	0.039	0.005	0.003	$\geq 6.6\%$	Manage 17th and 19th
24-Pulse									$< 5\%$	
AFE-PWM	0.037	0.005	0.001	0.019	0.022	0.015	0.004	0.0035	$< 5\%$	Generates EMI & RFI. Consider harmonics up to 99th

(4) 6-pulse drive with reactors and passive filters

(5) 6-pulse drive with active harmonic filters

(6) 12-pulse drive

(7) 12-pulse drive with reactors

(8) 12-pulse drive with passive filters

(9) 12-pulse drive with reactors and passive filters

(10) 18-pulse drive

(11) Active front end (AFE)

(12) PWM drives

7.1 SOLID-STATE DEVICES CARRIER FREQUENCY

1. SCRs (Silicon controlled rectifier): What is the rise time of the variable speed drive's output IGBTs? (a) 2 khz to 8 khz can operate in the 250 to 500 Hz range (4 to 8 times fundamentals).

2. BJTs (Bipolar junction transistor) can operate in the 1 to 2 kHZ range (16 to 32 times fundamentals).
3. Insulated gate bipolar transistors (IGBT) for the inverter section. IGBTs can turn on and off at a much higher frequency, 2 to 20 kHz (30 to 160 times fundamentals).
4. Typical AFE (Active front end with IGBT) switching frequency 3600 Hz (60 times fundamentals).

Table 7.5   Solid-State Device Sample Carrier Frequency

Solid-state device type	Carrier frequency	Explanation
SCR	250 Hz to 500 Hz	(a) 4 to 8 times the fundamentals
BJT	1 kHz to 2 kHz	(a) 8 to 16 times the fundamentals
IGBT	2 kHz to 20 kHz	(a) 16 to 160 times the fundamentals
PWM	2 kHz to 30 kHz	(a) 16 to 240 times the fundamentals

The motor electrical frequency is modulated at a much higher rate using pulse width modulation techniques at the range of 2 kHz to 30 kHz, which is called PWM carrier frequency. The carrier frequency at that range (2 kHz to 3 kHz) will have a very high dv/dt producing electromagnetic interference (EMI) or RFI, radio frequency interference.

Variable frequency drives generate harmonics creating many noise issues in the power system. Some of the noise implications are very complex. Therefore, it is very important to understand various noise issues, such as noise origination and effects, and then manage noise within a reasonable level. VFD-related noise can generate heat, degrading equipment and premature failures. Shipboard electrical equipment is usually designed for twenty to thirty years' life, depending on the application. However, due to VFD applications, the electrical equipment failure trend has changed, leading to premature failure.

Many propulsion auxiliaries are also being provided with VFD. Shipboard power distribution is mostly radial bus with some built-in isolation capability. The isolation is to ensure power segregation in case of major failures in one section of the switchboard. Therefore, VFD-related noise issues propagate the entire power system if there is no requirement nor any provision for system-level isolation.

Compared with other traditional propulsions, electrical propulsion with VFD is the most acceptable solution. Electrical propulsion also has a high power requirement due to propulsion power requirements, which can be from a few megawatts to 40 MW. The USCG Healy Ice breaker propulsion power is approximately 20 MW. Due to the propulsion power generation requirement, the power generation can be up to 15 kV. There are other smaller auxiliary propulsion requirements such as thrusters.

Harmonic calculation requirements are to establish a point of common coupling (PCC). For shipboard applications, the PCCs are the propulsion swbd bus and the power transformer secondary. The total harmonic distortion (THD) requirement is not to exceed 5%. The THD can be over 5% if the equipment at that system is selected to withstand the available harmonics. But in no case can the total harmonics be over 8%.

The following shipboard VFD applications

1. 18-pulse drive with phase shift transformers
2. 12-pulse drive with phase shift transformers and filters
3. 6-pulse drive with filters
4. Active front end drive
5. PWM drive

The 18-pulse drive is robust and expensive and demands space, due to three transformer requirements. However, the 18-pulse drive will meet harmonic requirements of THD 5%.

The 6-pulse drive is the least expensive, with minimum space requirements. However, the 6-pulse drive system may have a high THD level. Therefore, 6-pulse drives are used with filters to minimize the THD level to meet requirements. There are different types of filters available, such as passive filters and active filters. The 6-pulse drive with active filter and reactor has proven to be the best design in terms of size, cost, and reliability.

7.2 MIL-STD-1399 REQUIREMENTS

For shipboard power system MIL-STD -1399 power system harmonic distortion requirements:

Table 7.6   Harmonic Limits by MIL-STD-1399

Maximum total harmonic distortion	5%
Maximum single harmonic	3%
Maximum deviation factor	5%

7.3 IEEE STD 519 REQUIREMENTS (1992 AND 2014 VERSIONS)

For shipboard power generation and distribution-related harmonic management both voltage and current harmonic limits must be established. At various power distribution levels, both voltage and current harmonic levels must be established and then must be managed for system-level consideration. This can only be done when proper PCC is determined during the system design and development. The harmonic

Table 7.7   MV Power System Characteristics MIL-STD-1399

Characteristics	5 kV Class	8.7 kV Class	15 kV Class
Frequency			
1. Nominal frequency	60 Hz	60 Hz	60 Hz
2. Frequency tolerance	3%	3%	3%
3. Frequency modulation	0.50%	0.50%	0.50%
4. Frequency transient tolerance	± 4%	± 4%	± 4%
5. Worst-case frequency excursion from nominal frequency resulting from items 2, 3, and 4 combined, except under emergency conditions	± 5.5%	± 5.5%	± 5.5%
6. Recovery time from 4 or 5	2 sec	2 sec	2 sec
Voltage			
7. Nominal user voltage	4.16 kV rms	6.6 kV rms	11 kV rms
8. Line voltage unbalance	3%	3%	3%
9. User voltage tolerance			
9a. Average of three line-to-line voltages	± 5%	± 5%	± 5%
9b. Any one line-to-line voltage, including items 8 and 9a			
10. Voltage modulation	2%	2%	2%
11. Maximum departure voltage resulting from items 8, 9a, 9b, and 10 combined, except under transient or emergency conditions	± 6%	± 6%	± 6%
12. Voltage transient tolerance	± 16%	± 16%	± 16%
13. Worst-case voltage excursion from nominal user voltage resulting from items 8, 9a, 9b, and 10 combined except emergency conditions	± 20%	± 20%	± 20%
14. Recovery time from item 12 or 13	2 sec	2 sec	2 sec
15. Voltage spike	60 kV peak	75 kV peak	95 kV peak
Wave form (Voltage)			
16. Maximum total harmonic distortion	5%	5%	5%
17. Maximum single harmonic	3%	3%	3%
18. Maximum deviation factor	5%	5%	5%
Emergency Conditions			
19. Frequency excursion	(−100 to 12%)	(−100 to 12%)	(−100 to 12%)
20. Duration of frequency excursion	2 min	2 min	2 min
21. Voltage excursion	(−100 to 35%)	(−100 to 35%)	(−100 to 35%)
22. Duration of voltage excursion			
22a. Upper limit (+35%)	2 min	2 min	2 min
22b. Lower limit (−100%)	2 min	2 min	2 min

levels should be calculated at every PCC and then the accumulative effect should be established as required by MIL-STD-1399. If the harmonic levels are calculated to be higher than the requirement outlined in a dedicated system, the distribution at the coupling point must utilize proper equipment to withstand that particular harmonic level.

At the ship service user level such as 450 V and 120 V systems, the user contributes to the current and voltage harmonics, and measurements must be taken and appropriate measures taken to limit the propagation of harmonics. In general, each ship must be provided with calculated and measured harmonic levels at the time of delivery of the ship.

At the subsystem level, it is not allowed to control or alter the system impedance characteristics to reduce voltage distortion and passive equipment should not be added that affects the impedance characteristic in such a way that voltage distortions are increased. This adjustment is allowed only by the authority having jurisdiction for this action.

IEEE Std 519-1992 provides detailed understanding of harmonics, with examples to calculate harmonic levels. However, IEEE 519-1992 has been superseded with IEEE 519-2014, which provides some additional clarification, but gives no clarification of the requirements and no examples to follow. Therefore, the user of IEEE 519 for shipboard power system design and development must understand the differences between the two versions of IEEE 519. Some of the differences are given as follows.

7.3.1 Total Harmonic Distortion (THD)

Table 7.8   THD Definition Comparison (IEEE STD 519-1992 and 2014)

Total Harmonic Distortion (THD)		
THD IEEE 519-2014	TDD IEEE 519-1992	Remarks
TOTAL HARMONIC DISTORTION (THD): The ratio of the root mean square of the harmonic content, considering harmonic components up to the 50th order and specifically excluding inter-harmonics, expressed as a percent of the fundamental. Harmonic components of orders greater than 50 may be included when necessary. (IEEE 519-2014)	TOTAL HARMONIC DISTORTION (THD): This term has come into common usage to define either the voltage or current "distortion factor," where DISTORTION FACTOR is the ratio of the root mean square of the harmonic contents to the root mean square value of the fundamental quantity, expressed in percentage of fundamentals.	IEEE 519-1992 requirement was voltage or current. IEEE 519-2014 does not specifically state as voltage or current; instead, calls for harmonic contents.

7.3.2 Total Demand Distortion (TDD): Current Harmonics

Table 7.9 TDD Definition Comparison (IEEE 519-1992 and 2014)

Total Demand Distortion (TDD)		
TDD IEEE 519-2014	TDD IEEE 519-1992	REMARKS
The ratio of the root mean square of the harmonic content, considering harmonic components up to the 50th order and specifically excluding inter-harmonics, expressed as a percent of the maximum demand current. Harmonic components of orders greater than 50 may be included when necessary. (IEEE 519-2014)	TOTAL DEMAND DISTORTION (TDD): The total root-sum-square harmonic current distortion in percentage of the maximum demand load current (15- to 30-min demand). (IEEE 519-1992)	Current harmonic only.

TOTAL HARMONIC DISTORTION (THD): This term has come into common usage to define either the voltage or current "distortion factor," where DISTORTION FACTOR is: the ratio of the root mean square of the harmonic contents to the root mean square value of the fundamental quantity, expressed in percentage of fundamentals.

DISTORTION FACTOR (DF)

$$= \sqrt{\frac{Sum\ of\ the\ squares\ of\ amplitudes\ of\ the\ harmonics}{squares\ of\ the\ amplitudes\ of\ the\ fundamentals}} \times 100\%$$

(IEEE 519-1992)

TOTAL HARMONIC DISTORTION (THD): The ratio of the root mean square of the harmonic content, considering harmonic components up to the 50th order and specifically excluding inter-harmonics, expressed as a percent of the fundamental. Harmonic components of orders greater than 50 may be included when necessary. (IEEE 519-2014)

7.4 CALCULATE THE RMS HARMONIC VOLTAGE DUE TO THE RESPECTIVE HARMONIC CURRENT

The following method can be used: (Refer to ABS Guidance Notes for Control of Harmonics in Electrical Power Systems – Publication 150)

$$Vh = 3 \times I_h \times h \times Xgen \qquad (3.3)$$

Table 7.10   Total Harmonic Distortion (THD) Voltage Limit Comparison IEEE 519-1992 Version and IEEE 519-2014 Version)

Bus Voltage @ PCC	Individual Voltage Harmonic (%) per 519-2014	Individual Voltage Harmonic (%) per 519-1992	Total Voltage Harmonic Distortion-THD (%) per 519-2014	Total Voltage Harmonic Distortion = THD (%) per 519-1992	Remarks
V ≤ 1.0 kV	5	None	8	None	IEEE 519-2014 THD limit of 8% does not agree with established standards for shipboard power system.
1 kV < V ≤ 69 kV	3	3	5	5	Does agree with established standards for shipboard. For 519-1992 the voltage range was 69 kV and below power system.

where:

Vh   = (L-L) Line-to-line rms voltage of the harmonic order number "h"
Ih   = harmonic current at harmonic order number "h"
Xgen = generator reactance, in ohms
h   = harmonic number

To calculate the L − L rms harmonic voltage as a percentage of rms fundamental voltage

$$= \frac{Vh \times 100\%}{Vrms} \tag{3.4}$$

Example 3.47.1   Calculate the L-L rms 5th harmonic voltage for a 480 V generator with reactance X of 0.0346 ohms when the 5th harmonic current is 135 A. Also, express the harmonic voltage as a percentage of the fundamental rms voltage.

$$Vh = \sqrt{3} \times Ih \times h \times Xgen$$
$$= 1.732 \times 135 \times 5 \times 0.0346$$
$$= 40.45 \text{ V}$$

5th harmonic voltage as a percentage of the fundamental rms voltage:

$$= \frac{Vh \times 100\%}{Vrms} = = \frac{40.45 \times 100\%}{480} = 8.43\%$$

Note: Other percentage harmonic voltage components can be estimated by performing similar calculations for each harmonic current.

7.5 CURRENT HARMONIC MATTERS

Table 7.11 Total Demand Distortion (TDD) Current Limit (Extract from IEEE 519-2014 Table 2)

I_{SC}/I_L	$3 \le h < 11$	$11 \le h < 17$	$17 \le h < 23$	$23 \le h < 35$	$35 \le h < 50$	TDD	Remarks
< 20	4.0	2.0	1.5	0.6	0.3	5.0	See Note a
20 < 50	7.0	3.5	2.5	1.0	1.5	8.0	
50 < 100	10.0	4.5	4.0	1.5	0.7	12.0	
100 < 1000	12.0	5.5	5.0	2.0	1.0	15.0	

All power generation equipment is limited to these values of current distortion, regardless of actual I_{SC}/I_L

Where:

I_{SC} is maximum short circuit current at PCC

I_L is maximum demand load current (fundamental frequency component) at the PCC under normal load operating conditions

Note a: IEEE 519-2014 also gives allowance for high frequency current distortion limit if the 5th and 7th harmonic current is less than 1%. For a system with I_{SC}/I_L is less than 25 then all others may exceed up to a factor of 1.4. See IEEE 519-2014 Table 2 for details.

Table 7.12 Total Harmonic Distortion (THD) Voltage Limit Comparison between Different Drives

Attributes	6-Pulse	6-Pulse with Filter	12-Pulse	18-Pulse	Active Front End
Typical THD_V at the drive input terminal	@ 30%–40%	@5%–8%	5%–12%	5%–8%	3%–4%
Weight	100%	150%		240%	200%
Dimension	100%	140%		160%	170%

1. These are approximate values for information only. These values may widely vary from application to application.
2. Active front end unit is the most expensive one.
3. Drive transformers for 12-pulse and 18-pulse drives will have major weight and volume impact.

7.6 HARMONIC NUMBERING

Use this equation:

$$h = kq \pm 1$$

Table 7.13 Harmonic Contents for Various VFDS

Harmonic numbering	6-pulse drive	12-pulse drive	18-pulse drive	Remarks
3				Ignore triplen
5	5			
7	7			
9				Ignore triplen
11	11	11		
13	13	13		
15				Ignore triplen
17	17		17	
19	19		19	
21				Ignore triplen
23		23		
25		25		
27				
29				
31				
33				Ignore triplen
35		35	35	
37		37	37	
39				
41				
43				
45				Ignore triplen
47				
49			53	
51			55	

7.7 DNV REGULATION: HARMONIC DISTORTION

(a) Equipment producing transient voltage, frequency, and current variations shall not cause malfunction of other equipment on board, either by conduction, induction or radiation.

(b) In distribution systems the acceptance limits for voltage harmonic distortion shall correspond to IEC 61000-2-4 Class 2. (IEC 61000-2-4 Class 2 implies that the total voltage harmonic distortion shall not exceed 8%.) In addition, no single order harmonic shall exceed 5%.

(c) The total harmonic distortion may exceed the values given in (b) under the condition that all consumers and distribution equipment subjected to the increased distortion level have been designed to withstand the actual levels. The system and components ability to withstand the actual levels shall be documented.

(d) When filters are used for the limitation of harmonic distortion, special precautions shall be taken so that load shedding or tripping of consumers, or phase back of converters, do not cause transient voltages in the system in excess of the requirements. The generators shall operate within their design limits also with capacitive loading. The distribution system shall operate within its design limits, also when parts of the filters are tripped, or when the configuration of the system changes.

Guidance note:

The following effects should be considered when designing for higher harmonic distortion in (c):

- additional heat losses in machines, transformers, coils of switchgear, and control gear
- additional heat losses in capacitors for example in compensated fluorescent lighting
- resonance effects in the network
- functioning of instruments and control systems subjected to the distortion
- distortion of the accuracy of measuring instruments and protective gear (relays)

Remarks for DNV Harmonic requirements: Failure mode and effect analysis shall be performed at the shipboard system level for any filter application to ensure the cause and effect to the system as to the filter malfunction. If necessary, shipboard operational adjustment must be made to protect equipment and system.

7.8 EXAMPLES OF TYPICAL SHIPBOARD POWER SYSTEM HARMONIC CURRENT CALCULATIONS

Example 7.2   Power quality calculation: (Application Example for IEEE 519-1992 and IEEE 519-2014 compliance): Shipboard air compressor motor rated at 480 V, 100 HP VFD-fed from a 1500 kVA transformer of 4% impedance

Step 1: PCC default to be at VFD motor controller terminal

Step 2: Determine I_{sc} *(short circuit current)*. (Use transformer kVA rating and percent impedance)

$$I_{SC} = \frac{kVA \cdot 1000}{\sqrt{3} \cdot V_{LL} \cdot (\%Z/100)}$$

$$I_{SC} = \frac{1500 \cdot 1000}{\sqrt{3} \cdot 480 \cdot 0.04} = 45,105$$

Step 3: Determine I_L from the full load amp of the motor (124 AMP)

Step 4: Determine I_{SC} / I_L.

I_{SC}/I_L for the present case is $= 45,105/124 = 364$.

Table 10.2 (IEEE 519-1992) to determine limit - 15% A
TDD < 15% of Rated Fundamental Current $= 124 \times 0.15 = 18.6$ A
Therefore Ih $= 18.6$ AMP.

Example 7.3 480 V, 3-phase, 60 Hz, 200 hp air compressor onboard a ship is controlled by VFD drive. At the VFD (PCC) the short-circuit current is 50 KA.

— *full load current of the motor OF 200HP motor is 235 AMP*
- *The ratio of PCC short-circuit current and full load current (I_{SC} / I_L) is* = 50,000 / 235 = 213
- *According to IEEE 519-199) Table 10.2 the* **TDD** *for i_{sc} / i_l of 213 is 15%* **(TDD limit)**
- Therefore maximum calculated harmonic current is (213 × 15%) = 32 AMP

If the air compressor loading is above 117 AMP, then the system will not be in compliance with IEEE 519. Therefore, a harmonic management system such as the addition of a rector, passive harmonic filter, or active harmonic filter or combination thereof should be considered. Consideration of adapting any one of these harmonic management systems must be verified by proper calculations as well as the criticality of the load under consideration.

Example 7.4 For a shipboard power system at 690 V, for a 200 hp crane, at 156 Amp full load current is fed from a 2000 kVA transformer with 4.5% impedance. Calculate the ratio of the short-circuit current and the full load current.
The short-circuit current at the transformer:

$$I_{SC} = \frac{kVA \cdot 1000}{\sqrt{3} \cdot V_{LL} \cdot (\%Z/100)}$$

$$I_{SC} = (2000 \times 1000)/(1.732 \times 690 \times (4.5/100))$$
$$= 2000000/53.78 = 37,188 \text{ AMP}$$

The ratio of system short-circuit current to the full load current of the crane is I_{SC}/I_L = 37,188 / 156 = 238

Example 7.5 A 480 V system, 200 hp, 235 AMP crane motor is fed from a 2000 kVA transformer with 4.5% impedance. Determine how to comply with IEEE 519 if loaded at 125 AMP.
The short-circuit current:

$$I_{SC} = \frac{kVA \cdot 1000}{\sqrt{3} \cdot V_{LL} \cdot (\%Z/100)}$$

$$I_{SC} = (2000 \times 1000)/(1.732 \times 480 \times (4.5/100))$$
$$I_{SC} = 37,411 \text{ AMP}$$

The ratio between the short-circuit current and full load current:

$$I_{SC}/I_1 = 37,441/235 = 159$$

Per the IEEE 519-1992 look-up Table 10.2, the TDD limit for (I_{sc} / I_1) 159 is 15% of the operating current = (235 × .15) = 35.25 AMP. For example, the operating point of the pump is 125 AMP. At that operating point harmonic distortion is 25% of the operating current. (125 × 25%) = 31.25 AMP, which is less than 35.25 AMP.
(This meets the IEEE 519 requirements.)

7.9 CHOICE OF 18-PULSE DRIVE VERSUS 6-PULSE DRIVE WITH ACTIVE HARMONIC FILTER

Table 7.14   Filter Comparison Table

18-Pulse Drive	Active Harmonic Filter (AHF)	Remarks (Use of 6-Pulse Drive with AHF)
1. Will meet IEEE 519 harmonic requirements. The drive consists of three-phase shift transformers, transformer protection, and insulation heat management. The transformer may require pre-magnetization depending on the size. Transformer efficiency is low. For a large propulsion drive transformer, heat dissipation in the transformer room is very high. Transformer power cable primary and secondary adds a tremendous amount of weight. Shipboard transformer cable routing is a big challenge because so many cables are involved. Shipboard cable termination, particularly for VFD application, is also complex.	Will meet IEEE 519 harmonic requirements. The AHF can be integrated reactors to keep the size of the AHF smaller. The reactors are robust and long lasting without any maintenance.	Six pulse drive is such simpler than 18-pulse drive. 6-pulse drive, part replacement, repair, maintenance is less complicated than the 18-pulse drive.
2 Will not tolerate system voltage imbalance.	Will tolerate system voltage imbalance.	
3 Transformers are very heavy and space demanding.	AHF is not heavy, nor space demanding comparing with transformers.	
4 Transformer cooling is challenging leading to more cooling equipment demanding additional space and weight.	Do not know the cooling issues for AHF.	

(cont.)

Table 7.14   (*Cont.*)

18-Pulse Drive	Active Harmonic Filter (AHF)	Remarks (Use of 6-Pulse Drive with AHF)
	The AHF is a modularized design capable of multiple parallel connections. This contributes to easier size selection for ships due to headroom limitations. Due to the application of intelligent electronics and IGBTs, the system can deliver corrective amount of current and manage harmonics at the acceptable level.	
The transformer cost will increase due to the increase in the cost of copper.	The AHF cost will come down as the power electronic use increases and the rating increases.	

7.10 TYPICAL SOFTWARE TO CALCULATE TOTAL HARMONIC DISTORTION AND FILTER APPLICATIONS

A typical shipboard power system with variable frequency drive (VFD) or adjustable speed drive (ASD) is shown in Figure 7.1, consisting of a 6600 V distribution bus, a 690 V distribution bus, and a 480V distribution bus. This example is shown to calculate the harmonic distortion level and then apply various harmonic mitigating features to reduce the harmonics to a level acceptable per IEEE 519.

To select the correct harmonic filters for the system shown in Figure 7.1, the two 690 V busses are analyzed separately. Additionally, since they are both main-tie-main systems, the system is modeled as two individual circuits, one on each side of the tie. The 690 V bus SWBD-1 is supplied by a 2500 kVA, 6.6 kV to 690 V transformer with 5.75% impedance and feeds (1) 300 kW 6-pulse VFD and (1) 900 kW active front end drive connected to 3 ea. 300 kW motors. The 6-pulse drive is assumed to have 4% internal DC impedance, typical for a variable frequency drive.

The Figure 7.2 is shown from TCI's HarmonicGuard® Solution Center (HGSC.TRANSCOIL.COM) with multiple sections. The software is intended to calculate harmonic contents and then provide solutions in compliance with IEEE 519 harmonic limits or any other desired limit. Using the "Solution Center," enter the transformer parameters into section 2 of the data entry form. In Section 3, select "Harmonic Correction" and enter the desired iTHD level. In Section 4, enter the appropriate motor data as shown in Figure 7.1.

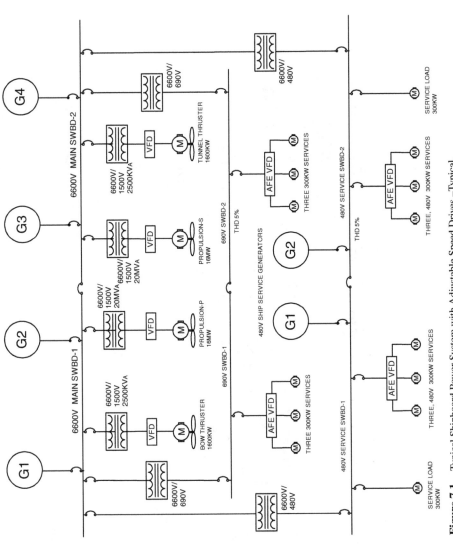

Figure 7.1 Typical Shipboard Power System with Adjustable Speed Drives –Typical.

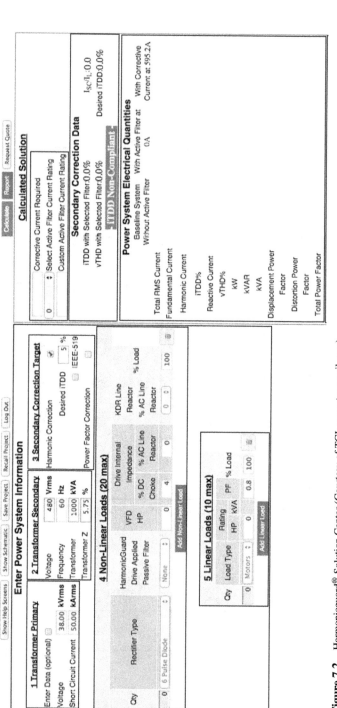

Figure 7.2 Harmonicguard® Solution Center (Courtesy of TCI; www.transcoil.com).

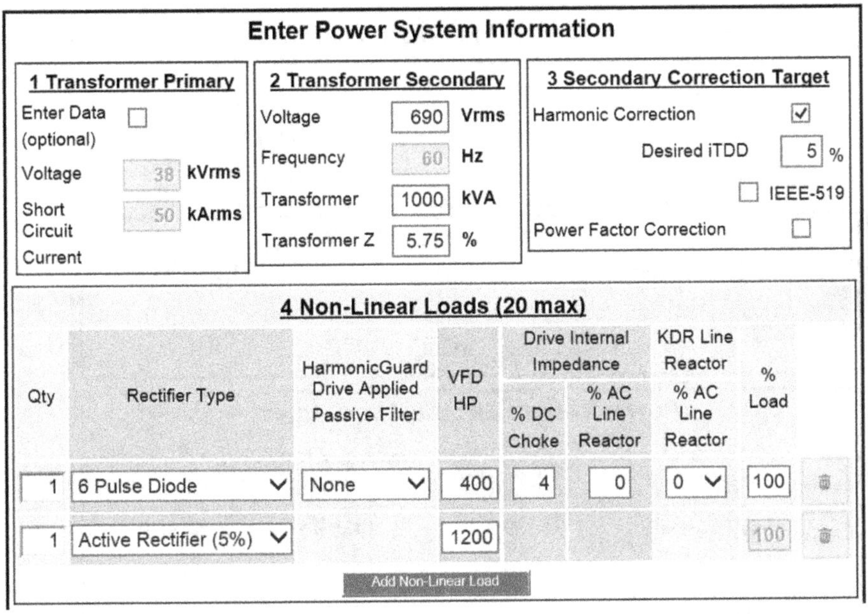

Enter Power System Information

1 Transformer Primary	**2 Transformer Secondary**	**3 Secondary Correction Target**
Enter Data ☐ (optional)	Voltage [690] **Vrms**	Harmonic Correction ☑
Voltage [38] **kVrms**	Frequency [60] **Hz**	Desired iTDD [5] %
Short Circuit Current [50] **kArms**	Transformer [1000] **kVA**	☐ IEEE-519
	Transformer Z [5.75] %	Power Factor Correction ☐

4 Non-Linear Loads (20 max)

Qty	Rectifier Type	HarmonicGuard Drive Applied Passive Filter	VFD HP	Drive Internal Impedance % DC Choke	% AC Line Reactor	KDR Line Reactor % AC Line Reactor	% Load	
1	6 Pulse Diode ∨	None ∨	400	4	0	0 ∨	100	🗑
1	Active Rectifier (5%) ∨		1200				100	🗑

Add Non-Linear Load

Figure 7.3 Typical Data Entry for Harminc Management Solutions.

In section 5, enter the rating of the across-the-line motor (Dedicated Load – 1).

In the upper right hand corner of the form, select the "Calculate" button. Using these parameters, 152.0 AMP of corrective current are required to achieve the desired 5% iTHD. In this case a 200 AMP active harmonic filter can be used; however, adding additional external impedance in front of the 6-pulse VFD may significantly reduce the required corrective current.

Using the drop-down box under Line Reactor, select a 3% line reactor. Re-evaluating the system with this additional impedance shows that the required corrective current is now 74.6 AMP with the additional drive impedance. Using the drop-down box under the Calculated select a 100A active filter and press the calculate button. In this case the HarmonicGuard Solution Center shows the following results:

7.11 HARMONIC RECOMMENDATIONS (IEEE 45.1 PARTIAL EXTRACT)

It is the cognizant design engineer's responsibility to be concerned about harmonic distortion issues and interface with the appropriate component vendors, to minimize risk, schedule and cost related to possible damage from harmonics. Harmonic distortion is more prevalent in modern power systems that employ large non-linear (non-sinusoidal) loads (e.g., power electronic power conversion equipment using switching electronics). These non-linear loads produce harmonics which can compromise

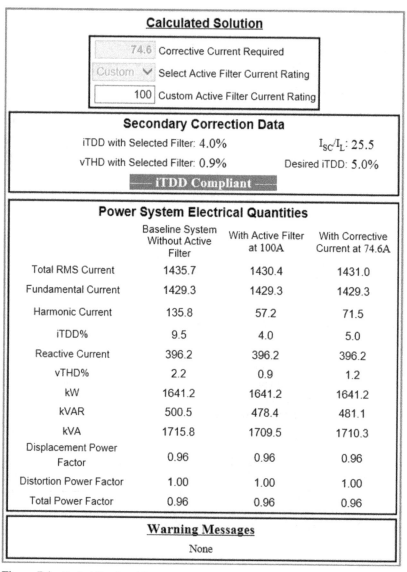

Figure 7.4 Typical Results of Harmonic Management with Active Filter and Reactor Option.

reliability of the power system. This is because harmonics take energy from the fundamental wave form and distribute that energy into other waveforms which are known as harmonics and sub harmonics which cannot be used by the loads. In doing so, the energy taken from the fundamental waveform is lost. That energy is displaced into other harmonic waveforms, is not used by loads and adds to heat losses in generation, transmission, transformation and motors. The resultant (unwanted) harmonic currents are ultimately sourced from the operating power generation equipment. In addition to heating losses, harmonic currents can increase structure-borne vibrations in connected loads and generators and introduce voltage distortion throughout the power system.

The design engineer needs to understand that the actual power factor for the power system reflects the vector sum of the fundamental waveform and all of the harmonic waveforms. Non-linear loads, such as rectifier circuits, don't typically shift the current waveform, rather they distort it. These distorted waveforms contain harmonic components which do no useful work and therefore are reactive in nature. For nonlinear loads, the power relationship becomes the vector sum of real power (kW), inductive or capacitive reactive power (kVAR) and distortion reactive power (also kVAR) to produce the apparent power which the power system must deliver (kVA). Power factor by definition is the ratio of kW to kVA and for nonlinear loads, the kVA includes a harmonic component. True power factor (TPF) becomes the combination of displacement power factor (dPF) and distortion power factor (hPF). Displacement PF is still equal to CosΦ, with Φ being the angle between the fundamental current and voltage. Displacement PF can be either leading or lagging. Distortion PF is then True PF (kW/kVA) divided by the dPF. Distortion PF is neither leading nor lagging.

Current harmonics in an AC power system generate voltage distortion as they flow through system impedances which can cause:

(a) Additional heating of power system components

(b) Inductive interference with communication, control systems and electronics

(c) Over current in capacitors

(d) Generator excitation or AVR control problems

(e) Resonance with power system impedance which amplifies problems

(f) Unwanted mechanical vibrations

On marine applications with generator supplied systems, the higher source impedance can result in much higher levels of voltage distortion when nonlinear loads are present. This is why all marine certifying bodies, including ABS, DNV GL, Lloyds, etc., have mandatory harmonic voltage distortion limits. It is recommended that the design engineer consider the most severe of these standards when addressing harmonics.

Equipment that can cause harmonic distortion includes AC or DC adjustable speed drives (ASDs), soft starters, power supplies, silicon controlled rectifiers (SCR) and solid-state frequency and voltage converters. Power electronics equipment is the source of electromagnetic interference in electrical power systems. Effect of high harmonics analysis should be based on the highest switching frequency in power electronics equipment. The gate current rise-time in power semiconductors should be limited as much as practical without significant reduction in the power conversion efficiency. The power electronics design should meet EMI requirements of IEC 61000 or MIL-STD.MIL-STD-461E-1999.

The fundamental waveform non-linear distortion, where current draw is dissimilar from the applied voltage, causes harmonics. These harmonics are "steady-state" (continuous) and therefore are power quality concern. Power system harmonic analysis also needs to evaluate "line notching" and possible ringing. Line notching introduces non-characteristic rectifier harmonics and should be addressed during the design process harmonic analysis. Line notching is normally caused by SCRs, thyristors, and diode rectifiers (to a lesser degree) when the line current commutates or transfers from one phase to another. During the commutation period, the two phases are short-circuited because they are electrically connected together for very short time durations through the converter bridge and the ac source impedance. The end result is voltage drops towards the zero crossing as the current increases, limited only by the circuit impedance. Line notching can seriously impact the system power quality especially when resonance within the power system causes ringing.

The power system design engineer needs to understand that firmware within power electronic power converters sense the input power waveform to rectify the AC power while minimizing line notching. If the source provides three wire isolated power then the internal sensing for the firmware needs to be from line- to-line with an algorithm to calculate the zero crossing for minimal line notching. If the source provides an isolated or grounded neutral with four wires then the internal sensing can be from line-to-line with an algorithm to calculate the zero crossing or line-to-ground without an algorithm for minimal line notching.

The degree of power factor (phase control) will affect the magnitude of the harmonics on AC power systems. Generators use voltage regulators and governors to control voltage and frequency that enable parallel operation. The combination of harmonic distortion and line notching can cause the generators to hunt resulting in voltage and frequency regulation instability within the control loops of the voltage regulators and governors. This becomes even more severe when generators with excessively high source impedance are specified in order to reduce the power system fault level.

Harmonic mitigating options:

(a) AC or DC reactors: Reactors are relatively easy to apply and will typically lower the current distortion drawn by the ASD or other nonlinear device by approximately 50% but this is very often not enough to meet acceptable voltage distortion limits. Typical values of reactance used are 3% to 5%. Simply increasing the impedance of the reactor to further reduce current harmonics will have minimal effect and can lead to excessive voltage drops which will reduce the output power capability of the ASD.

(b) Multipulse ASDs: 12, 18, 24 or higher pulse level ASDs are available with harmonic current reduction increasing with the pulse number. Phase shifting transformers or autotransformers are either built into the ASD or supplied separately. These transformers will add losses reducing the efficiency of the ASD and can significantly increase the space requirements. Also, the

effectiveness of the phase shifting in cancelling harmonics is very suscepti-
ble to background voltage distortion and voltage imbalance. As little as 2%
imbalance can drop the performance to levels no better than a 6-pulse ASD
with AC or DC reactor.

(c) Tuned passive filters: Each parallel connected tuned passive filter will target
a single harmonic. Therefore, to address the most predominant harmonics,
multiple level filters are required. As a parallel connection, these devices
must be reviewed for suitableness whenever new loads are added or the
power system is modified. Under lightly loaded conditions, capacitive reactive power can be quite high so consideration must be given to the ability of
the generator's automatic voltage regulator (AVR) to handle this capacitive
reactance.

(d) Wide spectrum passive filter: These series connected low pass filters are
designed to reduce the full spectrum of characteristic harmonics drawn by
6-pulse ASDs. Some filters are capable of reducing current distortion levels to < 5% at full load. Some designs, but not all, introduce high levels
of capacitive reactance under lightly loaded conditions which could lead to
generator AVR operational issues. It is important to either select a filter with
low capacitive reactive power or include capacitor switching contactors.

(e) Parallel active harmonic filter (AHF): Parallel connected AHFs are designed
to provide the harmonic currents required by the connected nonlinear load.
If sized properly, reduction in current harmonic distortion can be quite significant at the targeted harmonics below the 50th provided all 6-pulse ASDs
are equipped with an AC reactor (at least 3%) or a DC choke. The AHF
accomplishes this by the use of IGBTs making it similar to the inverter of
an ASD. AHFs can inject higher frequency harmonics above the 50th, which
can cause problems at much lower distortion levels than the lower frequency
harmonics.

(f) Active Front-end ASD (AFE): AFE drives reduce input current harmonics
with the use of IGBTs to regulate the current drawn by the rectifier. Input
current distortion can be substantially reduced at harmonic levels below the
50th. However, as with the AHF, AFE drives also inject higher frequency
harmonics above the 50th raising the current total harmonic distortion levels above 5% when harmonics up to the 100th are taken into consideration.
These higher frequencies can cause problems at much lower distortion levels than the lower frequency harmonics. AFE drives can also control power
in both directions allowing regeneration of energy when that is a desired
function.

Harmonic frequencies and the resultant harmonic currents drive heat losses within
transformers. These losses include eddy currents, hysteresis, copper losses, and stray
flux losses, which can cause failure due to overheating. In addition, winding insulation
stress can result when high levels of dv/dt are present.

In extreme cases, there is the possibility of resonance between transformer winding inductance and supply capacitance that cause additional losses. Also, there is the potential for laminated core vibrations that can generate unwanted audible noise. To ensure power system robustness, transformers need to be de-rated, oversized or harmonic mitigating when supplying nonlinear loads if the current total harmonic distortion (ITHD) is expected to be excessive. "K-factor" calculations are a common method used when designing and selecting transformers for nonlinear loads. K-factor is a measure of the additional losses that harmonic currents create within the transformer.

IEEE/ANSI and IEC (IEEE Std. C57.12.00, C57.12.90, or IEC 60076-1) have standards to de-rate transformers and it is the responsibility of the cognizant engineer to determine the appropriate standard to use. Selecting a transformer with a K-factor rating no less than the K-factor rating of its load will ensure that the transformer will not overheat due to harmonics. This will however, have little or no effect in reducing the voltage distortion introduced by the transformer due to the harmonic voltage drops across its impedance. To reduce voltage distortion, harmonic mitigating transformers should be considered.

Similar to a transformer, losses due to harmonics and line notching will increase in rotating equipment. Additional copper and iron losses (eddy current and hysteresis) will appear in the stator windings, rotor circuit and rotor laminations. In addition, when fed from PWM Adjustable Speed Drives, high dv/dt from the inverter IGBT's can cause winding insulation stress.

Power cable losses also increase with harmonics, which needs to be taken into account when de-rating the cables. Also over-current protection devices, such as fuses and circuit breakers, must be selected carefully to prevent nuisance tripping or overheating.

When the nonlinear loads, such as converters, are small and in low numbers, they will typically not lead to serious harmonic problems. However, when AC to DC converter loads, such as motor drive front ends, become more significant, in relation to the power source, say 30% or higher, then special attention should be given to understand the impact on system harmonics and appropriate measures taken to reduce their negative effects.

7.12 HARMONIC SILENCING AND ARC PREVENTION (CURTSEY OF APPLIED ENERGY)

Shipboard power systems with propulsion VFD and all other VDFs generate transients, in addition to the transients generated by the arcing behavior of the electrical system. Protection systems as well as arc flash mitigating systems are available for ARC protection as well as personnel protection. There is a proven system available that prevents momentary system-level sag and manages generator speed as well as voltage regulation. The product is called "Phaseback".

Phaseback is KEMA tested and very reliable for this application. Phaseback operates instantaneously and is automatic without any intervention.

Phaseback makes an instantaneous correction to any phase voltage that is out of balance. It prevents transient voltages such as those generated by switching transients or phase faults. Additionally, it utilizes the electromagnetic interaction of mutual induction to control voltages, protecting the device components and reducing degradation of the system and components.

Using electromagnetism in lieu of solid-state components, Phaseback reacts at the speed of the current flow at the beginning of a phase voltage imbalance. Energy is redirected from the phase with the higher voltages and fed back to the low-voltage phase. Phaseback uses inductive devices closely coupled for maximum efficiency and continuous duty. Phaseback is suitable for harsh environments as well as for shock and vibration.

Phaseback Eliminates Arc Flash Damage: Transient overvoltages and arcing ground-faults cause high frequency noise, insulation breakdown, control lockups, and premature equipment failure. Most solid grounds, which can cause an arc flash hazard, start as arcing ground faults. Phaseback prevents arcing ground faults, greatly reducing the potential for arc flash by over 85% and even in a scheduled or unscheduled power outage, Phaseback discharges the stored energy from the power distribution, reducing the arc-flash danger. Phaseback can be used in ungrounded and grounded systems up to 230 kV, and has been unaffected by any sized power surge or noise outside the system parameters.

Harmonic Reduction and Cancellation: Phase voltage harmonics cause eddy currents that cause heat in motors, transformers, and all other inductive devices. Phaseback reduces all harmonics, including zero, even, odd, and inter-harmonics, by 85% instantly.

Phaseback as a harmonic silencer reduces harmonic frequencies by the square law factor. The 3rd harmonic is filtered 9 times, the 5th harmonic is filtered 25 times and the 7th harmonic is filtered 49 times the typical loss of 60 Hz noise. This typically reduces all harmonics 50% to 85% from the 2nd harmonic through the 56,000,000th harmonic of 60 Hz.

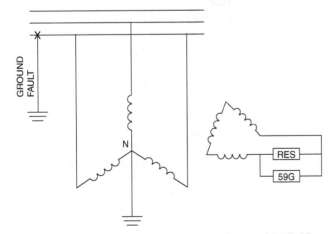

Figure 7.5   Phaseback Schematic Diagram -1 (Curtsey of Applied Energy).

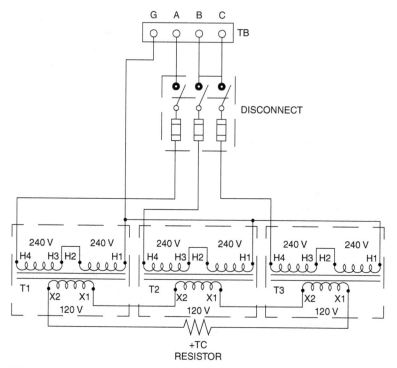

Figure 7.6   Phaseback Schematic Diagram-2 (Curtsey of Applied Energy).

The following is the explanation of the operation of the Phaseback unit: Ohms Law for AC circuits is Voltage equals Current multiplied by Ohms of Impedance, and, as the leading event, the voltage change offers the fastest reaction time when detected.

Phaseback operates very similar to a braking resistor, used to dissipate the charge current in the power system through the resistor in the series secondary circuit to lower the capacitive energy to a non-damaging level, thus mitigating the arc flash potential.

The series circuits in Phaseback have the same current through the entire circuit; circuits with equal ohms of impedance and equal current will maintain equal voltage. The transformers couple the primary and secondary circuits with magnetic lines of force. The secondary current of each transformer controls the primary current of that transformer and the voltage ratio is equal to the turn's ratio in the transformer; the current ratio is at the inverse of the voltage ratio in the transformer.

Phaseback uses (3) single-phase transformers to prevent core interaction. The center-tap or neutral of a wye-connected primary is connected to ground and the ungrounded single-phase secondary coils are connected in series with a large power resistor connected in series with those secondary coils.

The power resistor limits current to protect the transformers from overload and the current through and voltage across the resistor are in-phase (100% power factor). Harmonics are reduced and filtered through the resistor in the secondary as current circulates.

The transformers are selected to provide with equal secondary voltages and equal primary voltages. They are bi-directional, step-down and step-up types. The phase voltages across each primary coil are equal when the power system is balanced and cause the secondary voltages in Phaseback to be equal also. During a voltage imbalance on the distribution system, energy from the secondary is sent back to the low-voltage primary phase, and the positive feedback keeps the capacitive charge equal, causing the phase voltages on the power system to become stable.

Phaseback has two alarm points. The first point is located on the front of the panel; it is a ground fault indicating light. The second point is a secondary series circuit current detector with solid state (normally open) contacts that is factory set to alarm when 500 milliamps flow through the resistor. The early warning of the current detector provides maintenance the ability to correct the problem before a second fault forms, which can prevent multiple phases faulting through ground. This is a prime cause of arc flash.

For marine applications, a second current detector is connected to the ground connection. This current detector is connected to the neutral point of the primary connection in the unit for protection in the event that the ground cable is cut or otherwise disconnected.

Phaseback is the only transient voltage suppression system (TVSS) that can withstand the most powerful surges without damage to itself or anything it protects. Phaseback works on a fundamental electromagnetic principle that reacts at the speed of current flow.

The following are common power problems to consider related to power system abnormal behavior:

1. Transients
2. Interruptions
3. Sag / Undervoltage
4. Swell / Overvoltage
5. Frequency Variations
6. Harmonics

1. Transients:
An impulsive transient is a surge or a spike. Causes of impulsive transients include poor grounding, the switching of inductive load, fault clearing, and Electrostatic Discharge (ESD). The results can range from the loss or corruption of data to physical damage of equipment.

2. Interruptions:
An interruption is defined as the complete loss of supply voltage or load current. The causes of interruptions can vary but are usually the result of some type of electrical supply damage, such equipment failure, or a basic circuit breaker tripping.

3. Sag / Undervoltage:
A sag is a reduction of AC voltage at a given frequency for the duration of 0.5 cycles to 1 minute's time. Sags are usually caused by system faults and are also often the result of switching on loads with heavy startup currents.

4. Swell / Overvoltage:

A swell is the reverse form of a sag, having an increase in AC voltage for a duration of 0.5 cycles to 1 minute's time. For swells, high-impedance neutral connections, sudden (especially large) load reductions, and a single-phase fault on a three-phase system are common sources.

5. Frequency Variations:

There are all kinds of frequency issues from offsets, notching, harmonics, and inter-harmonics; but these are all conditions that occur largely in the end user's power system. These variations happen because harmonics from loads are more likely in smaller wye type systems.

Phaseback Design: Phaseback is a solution that focuses on prevention. The major difference in the design of Phaseback versus traditional surge protectors is that Phaseback corrects voltage in relation to ground rather than focusing on current.

By its patented design, Phaseback continuously stabilizes voltage relative to ground within a power system without sending imbalances to ground or by using solid-state technology such as metal oxide varistors (MOV). Phaseback systems are built to react at the speed of the infraction or at the speed of current flow which prevents power buildup and mitigates arc flashes. There is never any voltage leak through and the Phaseback unit will not degrade because of a transient event as do MOVs.

The components (matched three single-phase transformers) in this permanent solution are sized by the voltage class and kVA in which they will be employed. The voltage specification determines the appropriate turn ratios needed to properly size each system. All three transformers are spaced from one another by IEEE standards to prevent arcing or magnetic flux between each phase. Ohm's law explains how power reacts proportionately regardless of scope. Phaseback's effectiveness would be the same in a 300 kV system as it is in a 480 V system.

7.13 APPLICABLE POWER QUALITY STANDARDS INCLUDE

IEEE 519-1992: IEEE Recommended Practice and Requirements for Harmonic Control in Electric Power Systems

IEEE 519- 2014: IEEE Recommended Practice and Requirements for Harmonic Control in Electric Power Systems

IEC 61000-4-7 Voltage Flicker

IEC 61000-4-15 Harmonic Measurement

IEC 61000-4-30 Measurement Accuracy

ABS Guidance Notes for Control of Harmonics in Electrical Power Systems— Publication-150

Chapter 8

Shipboard Cable Application and Verification

8.0 INTRODUCTION: SHIPBOARD CABLE APPLICATION

Commercial shipboard electrical design and development-related cable construction requirements are in IEEE 1580 "Recommended Practice for Marine Cable for Use on Shipboard and Fixed or Floating Facilities," and cable installation requirements are in IEEE 45.8. US Navy cable construction and performance specifications are in MIL Specifications, such as MIL-C-24643, MIL-C-24640, MIL-C-915 as applicable. IEC cable requirements are in IEC standards such as 60092-350 series.

8.1 CABLE SIZE CALCULATION FUNDAMENTALS

For shipboard electrical system design and development, cable requirements must be examined to ensure that the cable selection is within the ampacity limit of the cable as well as within the allowable voltage drop limits for specific applications. For ASD applications, many constraints must be properly analyzed and proper cables must be selected. The following list shows some cable selection requirements:

- ABS allows cable selection for 100% full load amperes.
- USCG and ACP allow cable selection for 125% full load amperes.
- For IEEE cable use IEEE 45 ampacity only. Note that cable ampacity may be different from one manufacturer to another manufacturer.
- For IEC cable use, cable ampacity given in the IEC ampacity table should be used.
- The IEEE 45 requirement is to run cable in hangers and tiers, which allows uniform heat dissipation.
- If cable is pulled in an ambient other than 40 °C, i.e., the ambient of 45 °C or 50 °C, ampacity derating should be used as recommended.

Shipboard Power Systems Design and Verification Fundamentals, First Edition. Mohammed M. Islam.
© 2018 the Institute of Electrical and Electronics Engineers, Inc. Published 2018 by John Wiley & Sons, Inc.

- Consideration should be made when IEEE 45 cable is used for IEC equipment, given the fact that for IEC equipment all interconnecting dimensions are in millimeters.
- For overhead cable runs in locations such as auxiliary machinery space and under a continuous deck, proper ambient temperature must be considered for potential heat pockets.
- Fire integrity cable requirements also must be considered as applicable by regulation.
- Cables of different ambient ratings should not be run in the same wire-way. However, if the same wire-way is used, the lowest ambient rated ampacity should be used for all cables.
- Circuit breakers are usually tested in the factory for 40 °C ambient and tested with cable ambient ratings from 60 °C to 75 °C. If the circuit breakers are connected to higher ambient-rated cable, the circuit breakers may be subjected to higher temperature contributing to nuisance tripping, and malfunction.
- The circuit breaker and cable installation requirements should be thoroughly examined. If the feeder cable is directly connected to the circuit breaker lugs and ambient is restricted, the circuit breaker continuous rating must be evaluated. This is more so with the panel-mounted breakers, as the cable is directly connected to the lugs.
- Switchboard-mounted circuit breakers are not usually directly contacted with the cable as the cable gets connected to the bus. The bus spacing is usually engineered with sufficient spacing for heat dissipation. Therefore, switchboard circuit breaker derating requirement may be different from the panel board mounted circuit breakers.

8.2 SHIPBOARD CABLE FOR ASD AND VFD APPLICATIONS

(a) Cable for Adjustable Speed Drive (ASD) Shipboard Application and Verification

The adjustable speed drive (ASD) cable should be able to withstand operating conditions such as repetitive 1,600 volt peak voltage spikes from low-voltage (450 V) IGBT drives and at the same time not deteriorate the performance of other drive-system components.

Peak ASD voltage on a 450 V system can reach 1600 V, therefore the VFD cable should be rated at 2000 V.

8.3 CABLE REQUIREMENTS PER IEEE STD 45

IEEE Std 45-2002 Clause 23.2 Navy cable
Navy cables manufactured and tested in accordance with U.S. Navy military specifications MIL-C-24643A, MIL-C-24640A, or MIL-DTL-915 G may also be used,

provided they meet the IEEE Std 1202-1991 or CSA C22.2 No. 03, Vertical Flame Test.

IEEE STd 45-2002 Clause 23.3 Other shipboard cables
Marine shipboard cables meeting the performance requirements of IEEE Std 1580-2001 or UL 1309-1995, manufactured and tested to various national and international standards such as BS 6883-1999, CSA 22.2 No. 245, IEC 60092-350 series, JIS 3410, or other applicable marine standards are acceptable.

IEEE Std 45-2002 Clause 23.4 MI cable
MI cable as defined by NFPA 70-2002, NEC Article 332 may be used.

8.4 CABLE SHIELDING GUIDE PER IEEE STD 1143

(Extract from IEEE STD 1143-1994) IEEE Guide on Shielding Practice for Low Voltage Cables

Overall shield (Section 6.6.5)
An overall shield should be chosen for power surges, or other transients, and for protection against magnetic field (inductive) coupling.

Grounding and installation (Section 7): Introduction (Section 7.1)
Power grounding is for the purposes of electrical safety and to enhance the reliability and operation of electrically operated or supplied equipment within a facility. Signal grounding, on the other hand, is to assure "noise free" operation and reliability of the electronics system. Of greater importance, however, is the harmonizing of the safety and high-frequency grounding techniques so that electrical safety is not sacrificed in order to obtain satisfactory operation of the electronic system. The National Electrical Code (NEC) (ANSI/NFPA 70-1993) should be satisfied first; then, the signal grounding should be satisfied in a compatible way that does not undo the safety aspects of the installation. It is recommended that the system safety requirements be fully understood by all concerned with the design, and then the electronic requirements be overlaid onto the electrical safety basics, in such a way as to not diminish safety.

Shield grounded at one end (Section 7.2)
This is a technique to handle low-frequency noise and may not be appropriate for high-frequency noise or noise due to transients. The shielding may be grounded at either the sending end or the receiving end.

Shield grounded at sending end (Section 7.2.1)
Grounding the overall shield of a cable or a shielded pair at the sending end eliminates the transient voltage on the cable due to the electric field. However, a different condition prevails for the magnetically induced potential. Grounding the shield at the sending end has no effect on the magnetically induced component.

Shield grounded at receiving end (Section 7.2.2)
Grounding the overall shield or a shielded pair at the receiving end again prevents the electric field from reaching the cable, eliminating the electric field component. For the magnetically induced component, the input capacitance is now the only

circuit element between the voltage source and the cable. The capacitance between the cable and ground and the input capacitance forms a voltage divider. This arrangement reduces the surge voltage on the cable. The amount of reduction increases with cable length [B27].

Shield grounded at both ends (Section 7.3)

Here, for electric-field-induction, the displacement currents to the cable through capacitive coupling to the interference source are diverted to ground at each end and no transient voltage appears on the cable.

Grounding the shield at both ends completes a closed loop through the cable and ground mat system if the equipment or electronics at each grounded end are independently grounded. The magnetic field linking this loop induces a potential which in turn causes a secondary current to flow in the loop. The magnetic field due to this induced current opposes the primary field so that the net field in the loop is just sufficient to induce a potential drop related to the total resistive and reactive impedance around the current loop. This current flows axially along the shield of the cable.

Therefore, there is only a small magnetically induced voltage between the cable and the shield at the receiving end. With the shield grounded at two relatively greater in separation points, there is a risk that potential gradients in the ground mat system during faults may cause relatively larger shield currents to flow. Damage to the shielded cable may result if the shield is not a robust conductor. Therefore a heavier (thicker) shield (rather than a foil shield) is required. Use of an overall shield of corrugated 0.125 mm (0.005 in) copper or 0.2 mm (0.008 in) aluminum is usually sufficient to handle these surge currents [B4].

Transient protection with overall shields (Section 7.4)

When two electrical or electronic units are interconnected by a cable with the shield grounded at only one end, the load and source have an impedance to the common reference (usually ground). This establishes a "ground loop" between source and load, via the conductors, and the common "ground" reference. The common reference could be earth, and the example might represent wiring between two buildings. If a transient such as lightning causes fast changing electromagnetic lines of force to intercept the cable, then a current will be induced or coupled into all metallic portions of the cable except those open (ungrounded) at both ends, including the conductors and the shield(s).

The current induced into the inner conductors of the cable will be proportional to the loop impedance of each conductor, and it will be in the common-mode. This makes it difficult, for example, to detect and measure small differential currents with instrumentation connected line-to-line. In practice, the induced currents may be identical if they are common mode and very close in magnitude to one another because of similarities in loop impedances. The induced current is now capable of driving large surge currents into the source or load with the potential for destroying electronics connected to ends of the conductors.

A cable having an overall shield of sufficient thickness can be used to compensate for the induced currents in the conductors. For a cable with a shield that is effectively grounded at both ends, there is protection from the electromagnetic influence. (NOTE—Sometimes "bonded" is used to describe the grounding of the shield. Usually, the shield represents a lower total impedance throughout its length, and its impedance to the common grounding medium at either end is lower than that of the inner conductors. The shield must be grounded continuously over its length and at both ends so that there are no random paths established through the electronic components. With a lower loop impedance, the shield can develop a greater induced "noise" or surge current than the *inner* conductors. This condition is an advantage because as the shield carries more current, it also becomes another electromagnetic influence affecting the inner conductors. The result is a second induced current in the conductors from the current flow on shield. However, the induced current from the shield is opposite in phase to that induced by the original transient influence. Under these conditions, the inner conductors are confronted with two induced currents, but of opposite polarity. The result is a current in the conductor which is 180° out of phase with the noise current. As a result, the induced current from the shield cancels ("bucks out") the original surge ("noise") current in the conductors.

The transient magnetic fields from the interference current (I_m) induce a potential which causes transient current to flow in the shield (I_s) and conductors (I_c). Since the source of the shield current (I_s) and conductor current (I_c) was the same (i.e., the interference current), these currents are in phase. The shield and conductors are tightly coupled to each other in a parallel path over the entire length of cable. Therefore, the current flowing in the overall shield induces a second current (I_{sc}) into the conductors. I_{sc} is 180° out of phase with the shield current (I_s). As a consequence, this counter current nearly cancels out the interference current in the conductors. In order to handle the magnitude of current required to effectively cancel out the interference current, the cable needs an overall shield of 0.19 mm (7.5 mil) aluminum, 0.125 mm (5 mil) copper, or other overall shield or armor material. The canceling or bucking action of the counter current is never exact but does result in a definite and significant reduction in net induced current. This discussion has assumed that the voltage reference of the control circuit is an external ground. If the control circuit instrument is referenced to the shield, grounding the shield at both ends can increase the interference current in the conductors.

The effect of an overall shield consisting of longitudinally formed plastic coated 0.19 mm (7.5 mil) aluminum on reducing the effect of transients was verified experimentally. A test circuit of instrument cable with two pairs of twisted conductors was pulsed with a 62.5 kHz, 50 kV, 8730 A oscillatory surge. The voltages measured to ground of the overall shield and the conductors for an open or grounded shield are listed in table 7. This data shows that grounding the shield at both ends substantially reduces the induced voltage on the conductors. If the overall shield is left floating, or grounded only at one end, it will not be a complete circuit like the conductors. Therefore, counter current to cancel the interference current in the conductors cannot flow. As a result, the control circuit is not protected from the transient. The data also indicate that the use of a twisted pair, or foil shield, is not sufficient to shield signal conductors from electrical transient.

To maximize the flow of this counter current, everything that affects current flow should receive close attention. In particular, electrical continuity of the shield, electrically stable connections of low resistance, and grounds with low resistance are necessary. For two separated areas sharing a common grounding path, it is possible to drive a current directly through this ground path. Under this condition, the shield is still lower in impedance and forms a path in parallel with the signal conductors that usually have a higher impedance. Therefore, most of the noise current flows in the shield circuit rather than in the signal conductor.

Leaving the shield ungrounded ("floating the shield") will negate the ability of the shielding system to cancel electromagnetically induced currents. Overall shields grounded/bonded at both ends still retain their electrostatic shielding ability as well as provide electromagnetic shielding.

Grounding of cable with foil shields and overall shields (Section 7.5)
Grounding system (Section 7.5.1)
Grounding is applied first and foremost as a personnel safety protection device; second, as a means of limiting damage to equipment and cables; and third, for selective electrical system coordination (protection). The proper practice for modern plants is to provide single point grounding. A list of separate systems that may need to be connected to this ground is as follows:

(1) Power system ground (system ground)

(2) Instrument signal ground system

(3) Computer signal ground system

(4) Lightning protection ground system

Anywhere from one to four separate ground systems may be specified or designed. These systems use conductors to connect to each other and are grounded at one point to form a single point ground system.

The purpose of earth grounding is protection from lightning and transmission line ground faults, utilizing the earth as part of the return path for these currents. The grounding principle applicable for utilization voltages of < 600 V is not to use the "earth" as the current return path, but rather to use a system of interconnected conductors to equalize the voltage differences. This system can more effectively limit voltage differences than multiple ground connections.

For the transient and high-frequency interference control, the grounding system must be viewed as a possible interference distribution system. Transients injected onto the grounding system by power switching or other sources can propagate to all parts of the grounding conductors, including small-signal electronic circuits, that share the common grounding system. This can happen because the ground electrode has a surge impedance much greater than zero and the grounding conductors connected to it have surge impedances much less than infinity.

To prevent the grounding system from becoming an interference distribution system, the grounding must be coordinated with the shielding. That is, the local shield (equipment case, rack, shield room, or building) should serve as the grounding point for all circuits inside the shield, and grounding conductors should not penetrate any shield. This shielding and grounding topology is illustrated in figure 19, where the shield surfaces are shown as dashed lines and the grounding conductors are shown as solid lines. The shields are closed surfaces, and the grounding conductors are not allowed to penetrate the shield surfaces. The external ground is connected to the outside surface and the internal ground is connected to the inside surface. Thus, no transient or broadband interference waves are allowed to enter the protected space on the grounding conductor.

In particular, large transients generated outside the building (such as power switching transients or lightning) can be almost totally excluded from the sensitive integrated circuits inside the equipment case, since these waves are interrupted by several layers of shielding. Yet the safety requirements are satisfied, since within any shielded volume, all cases, racks, and equipment are grounded to the local shield structure. In addition, dc and 60 Hz fault currents can flow to actuate fuses and circuit breakers, since the shields are transparent to power frequencies.

The cable shield is a part of the shield system, it is not a grounding conductor. For a subsystem, the shield may consist of two equipment cases interconnected by a shielded cable. The complete shield (the two equipment shields and the cable shield) should be closed by connecting the cable shield to the equipment cases at each end—preferably with a circumferential bond between the cable shield and the equipment shield. The pigtail connection is adequate only for audio frequencies; at higher frequencies its inductive reactance is sufficient to produce a significant discontinuity between the shielded parts. The electrostatic shield is also a special case that cannot be applied to transient and broadband shielding.

Termination practice for overall shield (Section 7.5.3.1)
Shield bonding connectors are electrical shield terminating devices that serve a very important function in the protection of instrument and control systems from electromagnetic interference. The primary function of the connectors is to pass shield currents associated with electrical interference from the cable shield to the equipment shield without allowing the electrical interference to enter the equipment. Therefore, electrical stability of the connectors is essential.

An evolution in the design of connectors has resulted in connectors suitable for cable with plastic coated or bare overall shields. These connectors do not require the jacket to be stripped when the jacket is bonded adhesively to the overall shield. The connectors are slipped, over both the jacket and overall shield, using small tangs to lock into the jacket and to penetrate the plastic coatings on the inner side of the shield. Sophisticated test requirements have been developed by the telecommunications industry to assure reliable performance of the connectors. Generally, the following types of tests should be passed successfully for acceptance:

(1) Connector resistance

(2) Environmental requirements—vibration, temperature cycling, hydrogen sulfide exposure, and salt fog exposure

(3) Endurance tests—fault current and current surge

A common treatment of a shield at a connector is to insulate the shield with tape and connect it to the back shell of the equipment housing or case through a pigtail. The tail can introduce into the circuit an inductance of about 1000 nH/m, an inductance much higher than the transfer inductance produced by an equal length of overall shield. The shield current flowing through the inductance of the pigtail creates an interference voltage between the cable shield and the equipment case. It is important to keep grounding connections to the shield as short as possible.

For certain cables it may be possible to terminate the overall shield concentrically on the connector shell with no gaps in the circumference. When this procedure is followed, there is much less voltage introduced into the conductors for the following reasons:

(1) Length of the path through which the shield current must flow is shorter

(2) Field intensity inside the shield is nearly zero because the magnetic field is entirely external to the region occupied by the signal conductors (major factor)

(3) Field intensity external to the shield is reduced by virtue of the inherently larger diameter of the path upon which the current flows (minor factor).

8.5 CABLE: PHYSICAL CHARACTERISTICS

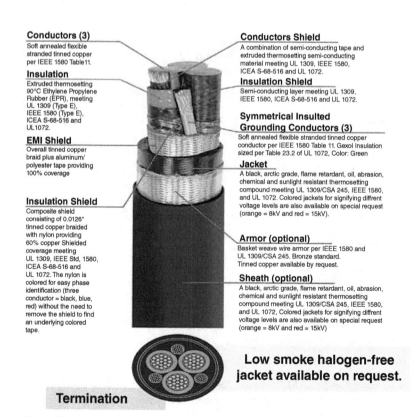

Conductors (3)
Soft annealed flexible stranded tinned copper per IEEE 1580 Table11.

Insulation
Extruded thermosetting 90°C Ethylene Propylene Rubber (EPR), meeting UL 1309 (Type E), IEEE 1580 (Type E), ICEA S-68-516 and UL1072.

EMI Shield
Overall tinned copper braid plus aluminum/ polyester tape providing 100% coverage

Insulation Shield
Composite shield consisting of 0.0126" tinned copper braided with nylon providing 60% copper Shielded coverage meeting UL 1309, IEEE Std, 1580, ICEA S-68-516 and UL 1072. The nylon is colored for easy phase identification (three conductor = black, blue, red) without the need to remove the shield to find an underlying colored tape.

Conductors Shield
A combination of semi-conducting tape and extruded thermosetting semi-conducting material meeting UL 1309, IEEE 1580, ICEA S-68-516 and UL 1072.

Insulation Shield
Semi-conducting layer meeting UL 1309, IEEE 1580, ICEA S-68-516 and UL 1072.

Symmetrical Insulted Grounding Conductors (3)
Soft annealed flexible stranded tinned copper conductor per IEEE 1580 Table 11. Gexol Insulation sized per Table 23.2 of UL 1072, Color: Green

Jacket
A black, arctic grade, flame retardant, oil, abrasion, chemical and sunlight resistant thermosetting compound meeting UL 1309/CSA 245, IEEE 1580, and UL 1072. Colored jackets for signifying diffrent voltage levels are also available on special request (orange = 8kV and red = 15kV).

Armor (optional)
Basket weave wire armor per IEEE 1580 and UL 1309/CSA 245. Bronze standard. Tinned copper available by request.

Sheath (optional)
A black, arctic grade, flame retardant, oil, abrasion, chemical and sunlight resistant thermosetting compound meeting UL 1309/CSA 245, IEEE 1580, and UL 1072, Colored jackets for signifying diffrent voltage levels are also available on special request (orange = 8kV and red = 15kV)

Low smoke halogen-free jacket available on request.

Termination

Figure 8.1   Typical MV Three Conductor Power Cable (Adapted).

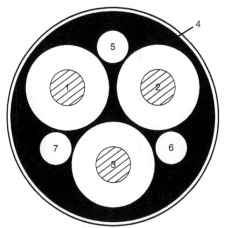

- Item 1: Basic Cable with conductor and insulation
- Item 2: Basic Cable with conductor and insulation
- Item 3: Basic Cable with conductor and insulation
- Item 4: Outer Overall Insulation
- Item 5: Interstice Ground Conductor-Bare
- Item 6: Interstice Ground Conductor-Bare
- Item 7: Interstice Ground Conductor-Bare

Figure 8.2   Typical Power System MV Three Conductor VFD Cable Details Without Armor (adapted).

Refer to NEC-2012 110.27, IEEE 1580, and IEEE 45.8

Shield Voltage: Table 6 – adjustments for ambient temperature. Note a) "Cable lengths should be limited to maintain a shield voltage below 25 V."

The new wording I propose is: "Cable installation should be designed to limit the shield voltage to less than 50 V AC at the maximum ampacity for the specific installation. If the voltage is greater than or equal to 50 V AC, then exposed shields shall be considered as live parts (in accordance with NEC-2012 110.27) and will require guarding or insulating for personnel protection and the cable manufacturer shall be consulted to determine the suitability of the cable construction to sustain excessive shield voltages."

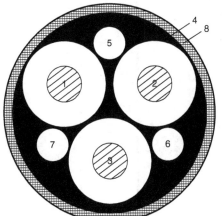

- Item 1: Basic Cable with conductor and insulation
- Item 2: Basic Cable with conductor and insulation
- Item 3: Basic Cable with conductor and insulation
- Item 4: Outer Overall Insulation
- Item 5: Interstice Ground Conductor-Bare
- Item 6: Interstice Ground Conductor-Bare
- Item 7: Interstice Ground Conductor-Bare
- Item 8: Armor

Figure 8.3   Typical Power System MV Single Conductor VFD Cable Details with Armor (Adapted).

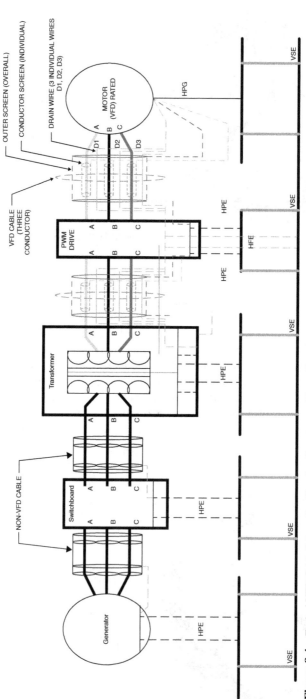

Figure 8.4 Typical VFD Cable Schematics Showing Generator to Motor (Sample-1).

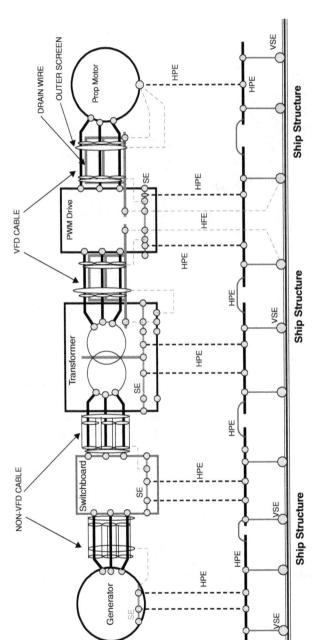

Figure 8.5 Typical Power System Cable (Sample-2).

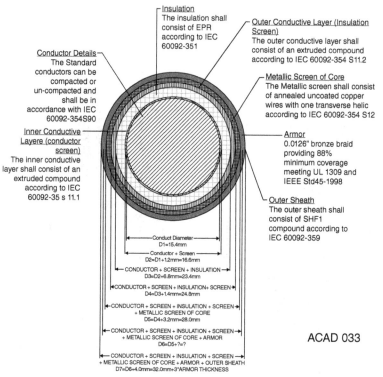

Figure 8.6   Typical Power System MV Single Conductor VFD Cable Details.

Though ferroresonance may be complex and hard to analyze, it need not be mysterious. Ferroresonance has been shown to be the result of specific circuit conditions, and can be induced predictably in the laboratory. Power system ferroresonance can lead to very dangerous and damaging overvoltages, but the condition can be mitigated or avoided by careful system design.

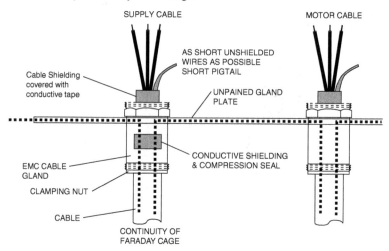

Figure 8.7   Typical Power Cable Deck Penetration.

8.6 CABLE INSULATION: TYPICAL

IEEE1580 Clause D.5.19.1.3 Insulation type

"E" Ethylene propylene rubber (EPR) (90 C)

"X" Crosslinked polyethylene (XPLE) (90C)

"P" Crosslinked polyolefin (100C)

"S" Silicone rubber (100C)

"LSX" Low smoke, halogen-free crosslinked polyolefin (90C)

"LSE" Low smoke, halogen-free ethylene propylene rubber (90C)

"T" Polyvinyl chloride (75C)

"T/N" Polyvinyl chloride/nylon (90C)

For insulation types E, X, T, T/N, and S where the VW-1 is the option, the letter "V" is added after the insulation type to indicate compliance with this optional requirement.

CABLE TYPE FOR POWER APPLICATION: (Design engineer must understand requirement differences among these cables so that the proper cable is used to meet operational requirements.)

(a) E type (EPR insulation) (90C)
  - STTB-4 = AWG 14 (4.11 kcmil), Single conductor, polyvinylchloride-insulated, thermoplastic polyvinyl chloride jacketed, and bronze armored
  - DTTA-4 = AWG 14 (4.11 kcmil), two-conductor, polyvinylchloride-insulated, thermoplastic polyvinyl chloride jacketed, and aluminum armored
  - TTTU-4 = AWG 14 (4.11 kcmil), three-conductor, polyvinylchloride-insulated, thermoplastic polyvinyl chloride jacketed, and without armor

(b) XPLE (X-cross-linked polyethylene insulated) (90C)
  - SXNU-250 = 250 kcmil, single conductor, cross-linked polyethylene insulated, Thermoset neoprene, without armor
  - DXNB-250 = 250 kcmil, two-conductor, cross-linked polyethylene insulated, Thermoset neoprene jacketed, and bronze armor
  - TXNA-250 = 250 kcmil, three-conductor, cross-linked polyethylene insulated, Thermoset neoprene jacketed, and aluminum armor

(c) P-type (Polyolefin insulated) (100C)
  - SPNBS-313 = Single conductor, 313 kcmil, polyolefin insulation with neoprene jacket, bronze armor, and overall sheath
  - DPNAS-313 = two-conductor, 313 kcmil, polyolefin insulation with neoprene jacket, aluminum armor, and overall sheath
  - TPNU-313 = three-conductor, 313 kcmil, polyolefin insulation with neoprene jacket, without armor

(d) "T" Polyvinyl chloride (75C)

- STNU-250 = 250 kcmil, single conductor, *Polyvinyl chloride* insulated, Thermoset neoprene, without armor
- DTNB-250 = 250 kcmil, two-conductor, *Polyvinyl chloride* insulated, Thermoset neoprene jacketed, and bronze armor
- TTNA-250 = 250 kcmil, three-conductor, *Polyvinyl chloride* insulated, Thermoset neoprene jacketed, and aluminum armor

(e) "T/N" Polyvinyl chloride/nylon (90C)

- ST/NNU-250 = 250 kcmil, single conductor, *Polyvinyl chloride/nylon* insulated, Thermoset neoprene, without armor
- DT/NNB-250 = 250 kcmil, two-conductor, *Polyvinyl chloride/nylon* insulated, Thermoset neoprene jacketed, and bronze armor
- TT/NNA-250 = 250 kcmil, three-conductor, *Polyvinyl chloride/nylon* insulated, Thermoset neoprene jacketed, and aluminum armor

IEC TYPE CABLE INFORMATION: (EXAMPLE ONLY TO DIFFERENTIATE BETWEEN IEEE-45, IEEE-1580 AND IEC CABLE IDENTIFICATION: IEC -60092-505 PART-350 SHIPBOARD POWER CABLECABLE SPEC

Table 8.1   Typical IEC Cable Construction Features

Cable Material Details	Insulation Type	Inner Covering and Sheath	Armor/ Screen	Outer Sheath	Remarks
FIRE RESISTANT TAPE PLUS INSULATION (HALOGEN FREE)	B				
ETHYLENE PROPYLENE RUBBER-EPR	R				
CROSSLINK POLYETHELYNE-XPLE	T				
THERMOPLASTIC COMPOUND-HALOGEN FREE	I				
HALOGEN FREE THERMOSET COMPOUND OR EVA	U				
INNER COVERING OR TAPPING (HALOGEN FREE)		F			
SCREEN (POSS WITH PE OR PP)		Y			
NO ARMOR			X		
COPPER WIRE BRAID (TINNED OR BARE)			O		
BRONZE WIRE BRAID			B		
GALVANIZED STEEL WIRE BRAID			C		
THERMOPLASTIC COMPOUND-HALOGEN FREE				I	
HALOGEN FREE THERMOSET COMPOUND OR EVA				U	
HALOGEN FREE MUD RESISTANT THERMOSET COMPOUND				B	

8.7 CABLE AMPACITY

IEEE STD 45-2002, TABLE-25 (CLAUSE 24.8), DISTRIBUTION, CONTROL AND SIGNAL CABLE

Table 8.25 Distribution, Control, and Signal Cables—Single-Banked, Maximum Current-Carrying Capacity (Types T, T/N, E, X, S, LSE, LSX, and P @ 45 °C Ambient)

AWG/ kcmil	Cross-sectional mm²	Circular Mils	Single-conductor T 75 °C	Single-conductor LSE LSX T/N E, X 90 °C	Single-conductor S, P 100 °C	Two-conductor T 75 °C	Two-conductor LSE LSX T/N E, X 90 °C	Two-conductor S, P 100 °C	Three-conductor T 75 °C	Three-conductor LSE LSX T/N E, X 90 °C	Three-conductor S, P 100 °C
20	0.517	1020	9	11	12	8	9	10	6	8	9
18	0.821	1620	13	15	16	11	13	14	9	11	12
16	1.31	2580	18	21	23	15	18	19	13	15	16
–	1.5	2960	20	24	26	17	20	22	14	17	18
15	1.65	3157	21	26	28	18	22	23	15	18	19
14	2.08	4110	28	34	37	24	29	31	20	24	25
12	3.3 1	6530	35	43	45	31	36	40	24	29	31
10	5.26	10,400	45	54	58	38	46	49	32	38	41
8	8.37	16,500	56	68	72	49	60	64	41	48	52
7	10.5	20,800	65	77	84	59	72	78	48	59	63
6	13.3	26,300	73	88	96	66	79	85	54	65	70
5	16.8	33,100	84	100	109	78	92	101	64	75	82
4	21.2	41,700	97	118	128	84	101	110	70	83	92
3	26.7	52,600	112	134	146	102	121	132	83	99	108
2	33.6	66,400	129	156	169	115	137	149	93	111	122
1	42.4	83,700	150	180	194	134	161	174	110	131	143
1/0	53.5	106,000	174	207	227	153	183	199	126	150	164
2/0	67.4	133,000	202	240	262	187	233	242	145	173	188
3/0	85.0	168,000	231	278	300	205	245	265	168	201	218
4/0	107.2	212,000	271	324	351	237	284	307	194	232	252
250 kcmil	126.7	250,000	300	359	389	264	316	344	217	259	282
262 kcmil	133.1	262,600	314	378	407	278	333	358	228	273	294
300 kcmil	152	300,000	345	412	449	296	354	385	242	290	316
313 kcmil	158.7	313,100	351	423	455	303	363	391	249	298	321
350 kcmil	177.3	350,000	372	446	485	324	387	421	265	317	344
373 kcmil	189.4	373,700	393	474	516	339	406	442	277	332	361
400 kcmil	203	400,000	410	489	533	351	419	455	286	342	371
444 kcmil	225.2	444,400	453	546	588	391	468	504	319	382	411
500 kcmil	253.3	500,000	469	560	609	401	479	520	329	393	428
535 kcmil	271.3	535,000	485	579	630	415	496	538	340	407	443
600 kcmil	304	600,000	521	623	678	450	539	585	368	440	478
646 kcmil	327.6	646,000	557	671	731	485	581	632	396	474	516
750 kcmil	380	750,000	605	723	786	503	602	656	413	494	537
777 kcmil	394.2	777,000	627	755	822	525	629	684	431	516	562
1000 kcmil	506.7	1,000,000	723	867	939	601	721	834	493	592	641
1111 kcmil	563.1	1,111,000	767	942	1025	637	784	854	523	644	701
1250 kcmil	635	1,250,000	824	990	1072	–	–	–	–	–	–
1500 kcmil	761	1,500,000	917	1100	1195	–	–	–	–	–	–
2000 kcmil	1013	2,000,000	1076	1292	1400	–	–	–	–	–	–

Ampacity adjustment factors for more than three conductors in a cable with no load diversity: Percent of values in Table 8.25 for Three-Conductor Cable as adjusted.

Number of Conductors for ambient temperature, if necessary

4 through 6: 80

7 through 9: 70

10 through 20: 50

21 through 30: 45

31 through 40: 40

41 through 60: 35

NOTES

(1) Current ratings are for AC or DC.

(2) For service voltage 1001 V to 5000 V, Type T, T/N, LSE, and LSX should not be used.

(3) Current-carrying capacity of four-conductor cables where one conductor does not act as a normal current-carrying conductor (e.g., grounded neutral or grounding conductor), is the same as three-conductor cables listed in Table 27 of IEEE STD 45-2002

(4) Table 8.25 is based on an ambient temperature of 45 °C and maximum conductor temperature not exceeding: 75 °C for type T insulated cables, 90 °C for types T/N, X, E, LSE, and LSX insulated cables, and 100 °C for types S and P insulated cables.

(5) If ambient temperatures differ from 45 °C, the values shown in Table 8.25 of IEEE STD 45-2002 should be multiplied by the following factors:

(6) The current-carrying capacities in this table are for marine installations with cables arranged in a single bank per hanger and are 85% of the ICEA calculated values [see Note 7]. Double banking of distribution-type cables should be avoided. For those instances where cable must be double banked, the current-carrying capacities in Table 8.25 should be multiplied by 0.8.

(7) The ICEA calculated current capacities of these cables are based on cables installed in free air, that is, at least one cable diameter spacing between adjacent cables. See IEEE Std 835-1994.

(8) For cables with maintained spacing of at least 1 cable diameter apart, the values from this table may be divided by 0.85.

Table 8.2   Low-Voltage Cable Ampacity Table

Ambient Temperature	Type T Insulated Cable	Type T/N, X, E, LSE, LSX Insulated Cable	Type S and P Insulated Cable
30 °C	1.22	1.15	1.13
40 °C	1.08	1.05	1.04
50 °C	0.91	0.94	0.95
60 °C	–	0.82	0.85
70 °C	–	0.74	0.74

Distribution cables—Single and Three Conductor single-banked, maximum current-carrying capacity (types T, T/N, E, X, S, LSE, LSX, and P @ 45 °C ambient)

Table 8.3   Low-Voltage Cable Ampacity Table with Metric Size Comparison and Ampacity Ratings

AWG/kcmil	Cross-sectional Converted mm²	Circular Mils	Three-conductor Ampacity T	LSE LSX T/N E, X	S, P	Metric Sizes in mm² STD-mm²	Single Cond Ampacity	3 Conductor Ampacity
			75 °C	90 °C	100 °C			
–	1.5	2960	14	17	18	1.0	19	
15	1.65	3157	15	18	19	1.5	21	15
14	2.08	4110	20	24	25	2.5	30	21
12	3.3 1	6530	24	29	31	4	40	29
10	5.26	10,400	32	38	41	6	51	36
8	8.37	16,500	41	48	52	10	71	50
7	10.5	20,800	48	59	63	None		
6	13.3	26,300	54	65	70	16	95	67
5	16.8	33,100	64	75	82	None		
4	21.2	41,700	70	83	92	25	125	89
3	26.7	52,600	83	99	108	None		
2	33.6	66,400	93	111	122	35	155	105
1	42.4	83,700	110	131	143	50	190	135
1/0	53.5	106,000	126	150	164	70	240	175
2/0	67.4	133,000	145	173	188	70	240	175
3/0	85.0	168,000	168	201	218	95	290	205
4/0	107.2	212,000	194	232	252	120	340	240
250 kcmil	126.7	250,000	217	259	282	120–150		
262 kcmil	133.1	262,600	228	273	294	None		
300 kcmil	152	300,000	242	290	316	150	385	270
313 kcmil	158.7	313,100	249	298	321	None		
350 kcmil	177.3	350,000	265	317	344	185	440	305
373 kcmil	189.4	373,700	277	332	361	None		
400 kcmil	203	400,000	286	342	371	185		
444 kcmil	225.2	444,400	319	382	411	185–240		
500 kcmil	253.3	500,000	329	393	428	240	520	365

Per IEEE 1580 Clause 5.4 Insulation shield (5-35 kV shielded cable)
Shielded cable rated 5-35 kV shall contain an insulation shield in accordance
with UL 1072.

The insulation shield compound should be free stripping from the underlying insulation. Minimum adhesion requirements, per UL 1072, shall be
maintained.

Corona discharge is caused by the intense electrical field surrounding the
conductors.

Nitrogen in the air space between conductors ionize in a discharge of energy.
Discharges harm the cable by degrading the insulation material and damaging the cable's shield. Corona discharges can also damage the drive electronics and cause power losses. A corona discharge can generate enough heat to
melt the conductor's insulation. Thermosetting insulations such as cross-linked
polyethylene (XPLE) provide much better resistance to corona discharges than
PVC.

**IEEE 45-2002-Table 29: Ampacity for Medium Voltage Power Cable,
Copper Conductor—Triplexed or Triangular Configuration (Single-Layered),
Maximum Current-Carrying Capacity Based on 45 °C Ambient—Single
Conductor**

Table 8.4   IEEE 45-2002-Table 29—Ampacity for Medium-Voltage Power Cable, Copper
Conductor—Triplexed or Triangular Configuration (Single-Layered), Maximum
Current-Carrying Capacity Based on 45 °C Ambient – Single Conductor

AWG/ kcmil	mm²	Circular Mils	Up to 8 kV Shielded	Up to 8 kV Shielded	8001- to 15 kV Shielded	8001- to 15 kV Shielded	8001- to 15 kV Shielded	8001- to 15 kV Shielded
			90 °C	105 °C	90 °C	105 °C	90 °C	105 °C
6	1330	26,240	92	106	–	–	–	–
4	2115	41,740	121	135	–	–	–	–
2	3362	66,360	159	187	164	187	–	–
1	4240	83,690	184	216	151	216	192	216
1/0	5350	105,600	212	245	217	242	220	245
2/0	6744	133,100	244	284	250	284	250	284
3/0	8502	167,800	281	327	288	327	288	327
4/0	10,720	211,600	325	375	332	375	332	375
250	12,670	250,000	360	413	366	413	366	413
263	13,310	262,600	371	425	377	425	376	425
313	15,860	313,100	413	473	418	471	416	471
350	17,730	350,000	444	508	448	505	446	505
373	18,930	373,700	460	526	464	523	462	523
444	22,520	444,400	510	581	514	580	512	580
500	25,330	500,000	549	625	554	625	551	625
535	27,120	5,353,000	570	648	574	648	570	648
646	32,750	646,400	635	720	638	720	632	720

Table 8.5 API-14F-Table 3—Ampacity for Three-Conductor Medium-Voltage Power Cable, 2001 V to 15 kV, Copper Conductor Single-Banked (Single-Layered), Maximum Current-Carrying Capacity Based on 45 °C Ambient

AWG/ kcmil	mm^2	Circular Mils	Up to 5 kV Nonshielded	Up to 8 kV Shielded	Up to 8 kV Shielded	8001- to 15 kV Shielded	8001- to 15 kV Shielded	8001- to 15 kV Shielded	8001- to 15 kV Shielded
			90 °C	90 °C	105 °C	90 °C	105 °C	90 °C	105 °C
8	8.37	16,510	48	–	–	–	–	–	–
6	1330	26,240	64	75	85	–	–	–	–
4	2115	41,740	84	99	112	–	–	–	–
2	3362	66,360	112	129	146	133	150	–	–
1	4240	83,690	130	149	168	151	170	149	172
1/0	5350	105,600	151	171	193	174	196	174	196
2/0	6744	133,100	174	197	222	199	225	198	225
3/0	8502	167,800	202	226	255	229	259	230	257
4/0	10,720	211,600	232	260	294	263	297	262	294
250	12,670	250,000	258	287	324	291	329	291	327
263	13,310	262,600	266	296	334	299	338	299	336
313	15,860	313,100	296	328	370	331	374	329	373
350	17,730	350,000	319	352	397	355	401	351	400
373	18,930	373,700	330	365	412	367	414	363	414
444	22,520	444,400	365	387	437	388	438	402	470
500	25,330	500,000	393	434	490	434	490	432	490
535	27,120	5,353,000	407	449	507	449	507	447	507
646	32,750	646,400	453	496	560	497	561	496	559
750	38,000	750,000	496	541	611	542	612	541	609

The column group headers read: Cross Sectional | Three-Conductor Cable | Three-Conductor Cable, with sub-headers: Up to 5 kV Nonshielded | Up to 8 kV Shielded | Up to 8 kV Shielded | 8001- to 15 kV Shielded (×4).

8.8 COMMERCIAL SHIPBOARD CABLE CIRCUIT DESIGNATION

For commercial shipbuilding, IEEE Std 45-2002, Appendix B provides circuit designations, to be used in electrical engineering drawing preparation. Table B1 gives traditional system designation prefixes and Table B2 provides system design details.

All electrical circuits, including those for power, lighting, interior communications, control, and electronics, are typically identified in the appropriate documentation and on-equipment labeling, such as nameplates, cable tags, wire markers, and so on, by a traditional system designation. These designations use a designation prefix, as follows.

Table 8.6   Typical Circuit Type, Circuit Designation, System Designation with Cable Types

Circuit Designation	Prefix	Circuit Number—Example Only	Explanation—Example Only
Ship service power (460 V)	P	P-412-T-250 (120 ft)	Power feeder, 460 V distribution, IEEE cable type T-250, three-conductor, 120 ft run
Emergency power (460 V)	EP	3EP-407-T-106 (50 ft)	
Propulsion power (6600 V)	PP	4PP-6612-T-26 (50 ft)	
Shore power (460 V)	SP	1SP-402-T-400 (125 ft)	
Lighting (120 V)	L	4L-114-T-212 (95 ft)	Lighting branch circuit, 120 V distribution, IEEE cable type T-212, three-conductor, 95 ft run
Emergency lighting (120 V)	EL	3EL-118-T-52 (60 ft)	
Interior communication	C		
Control	K		
Announcing: General	C-IMC	C-1MC-5	
Electric plant control and monitoring	C-DE	C-DE-7	
Emergency generator set control and indication	K-EG	K-EG-9	
Flooding alarm	C-FD	C-FD-32	
Radio antenna	R-RA	R-RA-5	
Radio satellite communication	R-RS		
Telephone: Automatic dial	C-J		
Telephone: Sound-powered (machinery control engineers)	C-2JV	C-2JV-15, TPS18-7 (55 ft)	Circuit 2JV, branch 15, 7 pairs of shielded twisted pairs. 18 AWG, 55 ft run

8.9 EXAMPLE 1: LOW-VOLTAGE 600 V/1000 V IEC CABLE DETAILS

Table 8.7   Example 1: Low-Voltage 600 V/1000 V IEC Cable Details

Cable Type	Explanation	Remarks
0.6/1KV-(LV) RFOU LV-(EPR)	INSULATION: (R): LOW-VOLTAGE, ETHYLENE PROPYLENE RUBBER-EPR, (F): INNER COVERING OR TAPPING (HALOGEN FREE): COPPER WIRE BRAID (TINNED OR BARE), (O): SCREEN – COPPER WIRE BRAID (TINNED OR BARE) (U): OUTER SHEATH-THERMOPLASTIC COMPOUND-HALOGEN FREE	LOW VOLTAGE 600 V/1000 V
0.6/1KV-(LV)RFBU LV-(EPR)	INSULATION: (R): LOW VOLTAGE, ETHYLENE PROPYLENE RUBBER-EPR, (F): INNER COVERING OR TAPPING (HALOGEN FREE): COPPER WIRE BRAID (TINNED OR BARE), (B): SCREEN – BRONE WIRE OR BRAID COPPER WIRE BRAID (TINNED OR BARE) (U): OUTER SHEATH-THERMOPLASTIC COMPOUND-HALOGEN FREE	LOW VOLTAGE 600 V/1000 V
0.6/1KV-(LV)RFCU LV-(EPR)	INSULATION: (R): LOW VOLTAGE, ETHYLENE PROPYLENE RUBBER-EPR, (F): INNER COVERING OR TAPPING (HALOGEN FREE): COPPER WIRE BRAID (TINNED OR BARE), (C): SCREEN – GALVANIZED STEEL WIRE BRAID (U): OUTER SHEATH-THERMOPLASTIC COMPOUND-HALOGEN FREE	LOW VOLTAGE 600 V/1000 V
0.6/1KV-(LV)TFOU LV-(XPLE)	INSULATION: (T): CROSSLINK POLYETHELYNE-XPLE (F): INNER COVERING OR TAPPING (HALOGEN FREE): COPPER WIRE BRAID (TINNED OR BARE), (O): SCREEN – COPPER WIRE BRAID (TINNED OR BARE) (U): OUTER SHEATH-THERMOPLASTIC COMPOUND-HALOGEN FREE	LOW VOLTAGE 600 V/1000 V

(Cont.)

Table 8.7   (*Cont.*)

Cable Type	Explanation	Remarks
0.6/1KV-(LV)TFBU LV-(XPLE)	INSULATION: (R): LOW VOLTAGE, ETHYLENE PROPYLENE RUBBER-EPR, (F): INNER COVERING OR TAPPING (HALOGEN FREE): COPPER WIRE BRAID (TINNED OR BARE), (B): SCREEN – BRONE WIRE OR BRAID COPPER WIRE BRAID (TINNED OR BARE) (U): OUTER SHEATH-THERMOPLASTIC COMPOUND-HALOGEN FREE	LOW VOLTAGE 600 V/1000 V
0.6/1KV-(LV)TFCU LV-(XPLE)	INSULATION: (R): LOW VOLTAGE, ETHYLENE PROPYLENE RUBBER-EPR, (F): INNER COVERING OR TAPPING (HALOGEN FREE): COPPER WIRE BRAID (TINNED OR BARE), (C): SCREEN – GALVANIZED STEEL WIRE BRAID (U): OUTER SHEATH-THERMOPLASTIC COMPOUND-HALOGEN FREE	LOW VOLTAGE 600 V/1000 V

8.10   EXAMPLE 2: MV VOLTAGE 8 KV/10 KV

Table 8.8   Example 2: Medium Voltage 600 V/1000 V IEC Cable Details

Cable Type	Explanation	Remarks
8 KV/10 KV-RFOU MV-(EPR)	INSULATION: (R): LOW-VOLTAGE, ETHYLENE PROPYLENE RUBBER-EPR, (F): INNER COVERING OR TAPPING (HALOGEN FREE): COPPER WIRE BRAID (TINNED OR BARE), (O): SCREEN – COPPER WIRE BRAID (TINNED OR BARE) (U): OUTER SHEATH-THERMOPLASTIC COMPOUND-HALOGEN FREE	8 KV/10 KV
8 KV/10 KV-RFOU MV-(EPR)	INSULATION: (R): LOW-VOLTAGE, ETHYLENE PROPYLENE RUBBER-EPR, (F): INNER COVERING OR TAPPING (HALOGEN FREE): COPPER WIRE BRAID (TINNED OR BARE),	8 KV/10 KV-RFOU

Table 8.8   (*Cont.*)

Cable Type	Explanation	Remarks
	(B): SCREEN – BRONE WIRE OR BRAID COPPER WIRE BRAID (TINNED OR BARE) (U): OUTER SHEATH-THERMOPLASTIC COMPOIND-HALOGEN FREE	
8 KV/10 KV-RFOU MV-(EPR)	INSULATION: (R): LOW-VOLTAGE, ETHYLENE PROPYLENE RUBBER-EPR, (F): INNER COVERING OR TAPPING (HALOGEN FREE): COPPER WIRE BRAID (TINNED OR BARE), (C): SCREEN – GALVANIZED STEEL WIRE BRAID (U): OUTER SHEATH-THERMOPLASTIC COMPOUND-HALOGEN FREE	8 KV/10 KV-RFOU
8 KV/10 KV-TFOU MV-(XPLE)	INSULATION: (T): CROSSLINK POLYETHELYNE-XPLE (F): INNER COVERING OR TAPPING (HALOGEN FREE): COPPER WIRE BRAID (TINNED OR BARE), (O): SCREEN – COPPER WIRE BRAID (TINNED OR BARE) (U): OUTER SHEATH-THERMOPLASTIC COMPOUND-HALOGEN FREE	8 KV/10 KV-TFOU
8 KV/10 KV-TFBU MV-(XPLE)	INSULATION: (R): LOW-VOLTAGE, ETHYLENE PROPYLENE RUBBER-EPR, (F): INNER COVERING OR TAPPING (HALOGEN FREE): COPPER WIRE BRAID (TINNED OR BARE), (B): SCREEN – BRONE WIRE OR BRAID COPPER WIRE BRAID (TINNED OR BARE)	8 KV/10 KV-TFBU
8 KV/10 KV-TFCU (MV-XPLE)	INSULATION: (R): LOW VOLTAGE, ETHYLENE PROPYLENE RUBBER-EPR, (F): INNER COVERING OR TAPPING (HALOGEN FREE): COPPER WIRE BRAID (TINNED OR BARE), (C): SCREEN – GALVANIZED STEEL WIRE BRAID (U): OUTER SHEATH-THERMOPLASTIC COMPOUND-HALOGEN FREE	8 KV/10 KV-TFCU

8.11 EXAMPLE 3: VFD CABLE LV (600 V/100) AND MV VOLTAGE (8 KV/10 KV)

	Epr – (R) Ethylene Propylene Rubber-Epr,	Xple – (T) Crosslink Polyethelyne-Xple
VFD - LV (600 V/1000 V) CABLE	0.6/1 KV-(LV) RFCU-VFD	0.6/1 KV-(LV) TFCU-VFD
VFD - MV (8 KV) CABLE	8 KV-(MV) RFCU-VFD	8 KV-(MV) TFCU-VFD

8.12 GROUND CONDUCTOR SIZE

IEEE 1580 TABLE 3 (GROUNDED CONDUCTOR FOR 2001 V OR LESS)
GROUNDING CONDUCTOR SIZE FOR CABLE RATED 2 KV OR LESS
FOR SINGLE RUN

Table 8.10 Grounding Conductor Size

			Grounding Conductor Size		
AWG/kcmil	Circular Mils	mm^2	75C Rated Conductor	90C Rated Conductor	100C Rated Conductor
20	1020	0.517	20	20	20
18	1620	0.821	18	18	18
16	2580	1.31	16	16	16
14	4110	2.08	14	14	14
12	6530	3.31	12	12	12
10	10,400	5.26	10	10	10
8	16,500	8.37	10	10	8
6	26,300	13.3	8	8	8
5	3300	16.8	8	8	8
4	41,700	21.2	8	6	6
3	52,600	26.7	6	6	6
2	66,400	33.6	6	6	6
1	83,700	42.4	6	6	6
1/0	106,000	53.5	6	6	6
2/0	133,000	67.4	6	4	4
3/0	168,000	85.0	4	4	4
4/0	212,000	107.2	4	4	4
250	250,000	126.7	4	3	3
262	262,000	133.1	4	4	4
313	313,000	158.7	3	3	3
350	350,000	177.3	3	3	3
373	373,000	189.4			
400	400,000	203			
444	444,000	225.2			
500	500,000	253.3			
535	535,000	271.3			
750	750,000	380			

GROUNDING CONDUCTOR SIZE FOR CABLE RATED ABOVE 2 KV FOR SINGLE RUN

8.13 DEVELOP MATH TO CALCULATE THE GROUND CONDUCTOR FOR PARALLEL RUN

GROUNDING CONDUCTOR SIZE FOR CABLE RATED ABOVE 2 KV FOR SINGLE RUN

Table 8.11   Grounding Conductor Size for Cable Rated Above 2 Kv for Parallel Run

| AWG/kcmil | Circular Mils | mm² | Grounding Conductor Size | | |
			75C Rated Conductor	90C Rated Conductor	100C Rated Conductor
20	1020	0.517	20	20	20
18	1620	0.821	18	18	18
16	2580	1.31	16	16	16
14	4110	2.08	14	14	14
12	6530	3.31	12	12	12
10	10,400	5.26	10	10	10
8	16,500	8.37	10	10	8
6	26,300	13.3	8	8	8
5	3300	16.8	8	8	8
4	41,700	21.2	8	6	6
3	52,600	26.7	6	6	6
2	66,400	33.6	6	6	6
1	83,700	42.4	6	6	6
1/0	106,000	53.5	6	6	6
2/0	133,000	67.4	6	4	4
3/0	168,000	85.0	4	4	4
4/0	212,000	107.2	4	4	4
250	250,000	126.7	4	3	3
262	262,000	133.1	4	4	4
313	313,000	158.7	3	3	3
350	350,000	177.3	3	3	3
373	373,000	189.4			
400	400,000	203			
444	444,000	225.2			
500	500,000	253.3			
535	535,000	271.3			
750	750,000	380			

8.14 CABLE DESIGNATION TYPE (TYPICAL SHIP SERVICE CABLE SYMBOL OR DESIGNATION)

Shipboard cable designations are standardized with unique symbols that directly correspond to the number of conductors constituting the cable. These typical cable symbols are listed for informational purposes. These abbreviations are often used to prepare shipboard working drawings such as one-line diagrams and power deck plans.

- "S" Single conductor distribution
- "D" Two-conductor distribution

- "T" Three-conductor distribution
- "F" Four-conductor distribution
- "Q" Five-conductor distribution
- "C" Control cable
- "tp" Twisted pair
- "TT" Twisted triad

8.15 CABLE COLOR CODE: SHIPBOARD COMMERCIAL CABLE

The cable color code was in IEEE Std 45-1998 Clause 8.31. This color code has been taken out from IEEE Std 45-2002 and can now be found in IEEE Std 1580-2001, Table 22. The color code is adapted from NEMA WC 57, Table E-1.

Table 8.12   Cable Color Code, IEEE 45 & IEEE-1580 (Extract)

Conductor Number	Base Color	Tracer Color	Tracer Color	Conductor Number	Base Color	Tracer Color	Tracer Color
1	Black			45	White	Black	Blue
2	White			46	Red	White	Blue
3	Red			47	Green	Orange	Red
4	Green			48	Blue	Red	Orange
5	Orange			49	Blue	Red	Orange
6	Blue			50	Black	Orange	Red
7	White	Black		51	White	Lack	Orange
8	Red	Black		52	Red	Orange	Black
9	Green	Black		53	Green	Red	Blue
10	Orange	Black		54	Orange	Black	Blue
11	Blue	Black		55	Blue	Black	Orange
12	Black	White		56	Black	Orange	Green
13	Red	White		57	White	Orange	Green
14	Green	White		58	Red	Orange	Green
15	Blue	White		59	Green	Black	Blue
16	Black	Red		60	Orange	Green	Blue
17	White	Red		61	Blue	Green	Orange
18	Orange	Red		62	Black	Red	Blue
19	Blue	Red		63	White	Orange	Blue
20	Red	Green		64	Red	Black	Blue
21	Orange	Green		65	Green	Orange	Blue
22	Black	White	Red	66	Orange	White	Red
23	White	Black	Red	67	Blue	White	Red
24	Red	Black	White	68	Black	Green	Blue
25	Green	Black	White	69	White	Green	Blue
26	Orange	Black	White	70	Red	Green	Blue

Table 8.12   (*Cont.*)

Conductor Number	Base Color	Tracer Color	Tracer Color	Conductor Number	Base Color	Tracer Color	Tracer Color
27	Blue	Black	White	71	Green	White	Red
28	Black	Red	Green	72	Orange	Red	Black
29	White	Red	Green	73	Blue	Red	Black
30	Red	Black	Green	74	Black	Orange	Blue
31	Green	Black	Orange	75	Red	Orange	Blue
32	Orange	Black	Green	76	Green	Red	Black
33	Blue	White	Orange	77	Orange	White	Green
34	Black	White	Orange	78	Blue	White	Green
35	White	Red	Orange	79	Red	White	Orange
36	Orange	White	Blue	80	Green	White	Orange
37	White	Red	Blue	81	Blue	Black	Green
38	Black	White	Green	82	Orange	Green	White
39	White	Black	Green	83	Green	Red	
40	Red	White	Green	84	Black	Green	
41	Green	White	Blue	85	Green5	White	
42	Orange	Red	Green	86	Blue	Green	
43	Blue Red	Red	Green	87	Black	Orange	
44	Black	White	Blue	88	White	Orange	

8.16 ASD (VFD) CABLE ISSUES FOR SHIPBOARD APPLICATION

The VFD cable should be able to withstand operating conditions such as repetitive 1,600 volt peak voltage spikes from low-voltage IGBT drives and at the same time not deteriorate the performance of other drive system components. Peak voltages on a 460 V system can reach 1200 V to 1600 V, causing rapid breakdown of motor insulation, leading to motor failure. If this is left uncontrolled, insulation failure may occur.

The same peak voltages that damage the motor can also damage the cable. In the perfect cable power delivery system the net instantaneous current flowing in the total cable system should be zero. This includes all phase conductors, all ground conductors, and shield. This can be achieved by a symmetrical cable.

Symmetrical cable

In a VFD installation the IGBT switches are in constant operation at high frequency, which produces an inverter output voltage with a PWM wave. This IGBT switching also causes a motor line to ground voltage, normally called a common mode voltage. Most AC drives, in addition to their normal three-phase output voltages, create a fourth unintended voltage to ground, known as common mode voltage.

Common mode current is current that leaves a source and does not come back to the source. In most closed-loop electrical circuits, most of the current returns to the source. However, there is a small amount of current in any circuit that is radiated and does not return.

Common mode voltages cause short high-frequency pulses of common mode current to flow in the safety earth circuits, and it is essential that the common mode currents return to the inverter without causing EMC-EMI problems in other equipment, which means that the common mode currents must not flow in the safety earthing (grounding) system.

The best and easiest way to do this is to use shielded VFD cables that are properly terminated and provide a low impedance path for common mode current to return to the inverter.

8.17 ABS STEEL VESSEL RULE: PART 4, CHAPTER 8, SECTION 4: SHIPBOARD CABLE APPLICATION

Single Conductor Cables

As far as possible, twin or multi-conductor cables are to be used in AC power distribution systems. However, where it is necessary to use single-conductor cables in circuits rated more than 20 A, arrangements are to be made to account for the harmful effect of electromagnetic induction as follows:

(i) The cable is to be supported on non-fragile insulators;

(ii) The cable armoring (to be nonmagnetic, see 4-8-3/9.11) or any metallic protection (nonmagnetic) is to be earthed at mid span or supply end only;

(iii) There are to be no magnetic circuits around individual cables and no magnetic materials between cables installed as a group; and

(iv) As far as practicable, cables for three-phase distribution are to be installed in groups, and each group is to comprise cables of the three phases (360 electrical degrees). Cables with runs of 30 m (100 ft) or longer and having a cross-sectional area of 185 mm^2 (365,005 circ. mils) or more are to be transposed throughout the length at intervals of not exceeding 15 m (50 ft) in order to equalize to some degree the impedance of the three phase circuits. Alternatively, such cables are to be installed in trefoil formation.

ASD (VFD) Cables selection criteria:

Cable used to connect the VFD drive to the motor affect how well the VFD system handles the phenomena with potential to disrupt operations. Cables designed specifically for VFD applications (see Figure 1) withstand these phenomena. And, equally important, they do nothing to make the situation worse. A generic motor supply cable is not designed with the specific requirements of VFD system in mind actually cause problems.

Here are some of the things to look for in a VFD cable.

Insulation

The insulation must have excellent dielectric properties to prevent breakdown from the stresses of voltage and current spike, corona discharges, and so forth.

Cross-linked polyethylene (XLPE) has the required properties, making it a better choice than standard polyethylene or PVC.

XLPE insulation helps reduce the potential effects of harmonics and corona discharge, therefore providing stable electrical performance that prolongs the life of the cable. This reduces the need for costly downtime due to cable failure.

Shielding

Shielding serves the double purpose of keeping noise generate by the VFD cable from escaping and preventing noise generated outside the system from being picked up. There are three types of shields typically used:

(1) Foil shield

Foil shielding uses a thin layer of aluminum, typically laminated to a substrate such as polyester to add strength and ruggedness. It provides 100% coverage of the conductors it surrounds, which is good. It is thin, which helps keep cable diameters small, but is it harder to work with, especially when applying a connector.

(2) Woven braid

A woven braid provides a low-resistance path to ground and is much easier to termination by crimping or soldering when attaching a connector. Depending on the tightness of the weave, braids typically provide between 70% and 95% coverage. Because copper has higher conductivity than aluminum and the braid has more bulk for conducting noise, the braid is more effective as a shield. But it adds size and cost to the cable.

(3) Both foil and braid shields

A third approach combines both foil and braid shields in protecting the cable. Each supports the other, overcoming the limitations of one with its own compensating strengths. As shown in Figure 8.2, this presents shielding effectiveness superior to either approach alone.

One purpose of the shield is to provide a low-impedance path to the ground. Any noise on the shield thus passes to the ground. A poor ground connection can increase the impedance, which also increases the potential for noise to be coupled to nearby cables or equipment.

Cable Geometry

A round, symmetrical cable gives the best electrical performance and resistance to deleterious effects. One reason is that round symmetry creates more uniform electrical characteristics in the cable. The electrical characteristic of one conductor doesn't differ significantly from another. For example, having the ground wires placed symmetrically around the cable provides a more balanced grounding system.

The Right Cable Prevents Problems

The cable selected for interconnecting a variable-frequency drive to a motor can significantly influence the reliability of the system.

The wrong cable can increase problems and lead to premature failure.

The right cable, on the other hand, can actually reduce the potential from problems like corona discharge or standing waves.

Alpha Wire offers a broad line of cables specifically designed to allow a VFD system to operate flawlessly to its fullest potential.

DRAIN WIRE:

The effect of an out-of-balance three-phase load causing the neutral conductor to become a heat emitter should not cause any overheating to the group of three-phase four-wire cables. However, when harmonics are present, as shown in Figure 8.1(b), the line current in the three phases are no longer balanced sine waves, and if they are in triplen order (i.e., 3n), they will be additive in the neutral. Now the neutral conductor becomes a fourth and additional current-carrying conductor. As a result, it is an additional heat-emitting source in the group of four conductors.

In view of the fact that the neutral conductor is now a current-carrying conductor, hence a heat emitter, steps need to be taken to take account of the extra heat that is produced by the neutral conductor in a three-phase circuit. The update published by the IEE (now the IET) [13] states that for every 8 °C increase above the maximum core conductor the continuous operating temperature the life of the cable will be halved (e.g., 25 years reduced to 12.5 years). A method is thus required to size the cable accordingly to dissipate the extra heat that is being generated within a group of three-phase four-wire conductors to ensure that the group of four conductors does not

8.18 GROUNDING CONDUCTOR SIZE: FOR CABLE RATED 2 KV OR LESS FOR SINGLE RUN

Table 8.13   Grounding Conductor Size for 2 KV or Less

AWG/kcmil	Circular Mils	mm²	Grounding Conductor Size		
			75C Rated Conductor	90C Rated Conductor	100C Rated Conductor
20	1020	0.517	20	20	20
18	1620	0.821	18	18	18
16	2580	1.31	16	16	16
14	4110	2.08	14	14	14
12	6530	3.31	12	12	12
10	10,400	5.26	10	10	10
8	16,500	8.37	10	10	8
6	26,300	13.3	8	8	8
5	3300	16.8	8	8	8
4	41,700	21.2	8	6	6
3	52,600	26.7	6	6	6
2	66,400	33.6	6	6	6
1	83,700	42.4	6	6	6
1/0	106,000	53.5	6	6	6
2/0	133,000	67.4	6	4	4
3/0	168,000	85.0	4	4	4
4/0	212,000	107.2	4	4	4
250	250,000	126.7	4	3	3
262	262,000	133.1	4	4	4
313	313,000	158.7	3	3	3
350	350,000	177.3	3	3	3

Using Separate Neutral Conductors

– On three-phase branch circuits, another philosophy is to not combine neutrals, but to run separate neutral conductors for each phase conductor. This increases the copper use by 33%. While this successfully eliminates the addition of the harmonic currents on the branch circuit neutrals, the panel board neutral bus and feeder neutral conductor still must be oversized.

Chapter 9

Grounding, Insulation Monitoring Design, and Verification

9.0 INTRODUCTION

Shipboard power generation and distribution is mainly a three-phase three-wire ungrounded system. In general, the ungrounded system,does not establish ground reference through the metallic ship hull or the electric ground bus or the electric ground conductor collecting the ship's ground plane.

The shipboard ungrounded electrical distribution system has capacitive ground through system-level capacitance.

An ungrounded electrical system is used on shipboard for the following reasons:

(a) To remain operational during single-phase fault

(b) To prevent contact with a dangerous voltage if electrical insulation fails or an electric power system fault occurs.

(c) To reduce the arc or flash hazard to personnel who may have accidently caused or who happen to be in close proximity to the ground fault.

(d) To minimize equipment damage from fault overvoltage, such as switchgear, drives, transformers, cable, and rotating machines

(e) To reduce static electricity build-up on rotating machines

(f) To reduce common mode transient overvoltage

(g) To eliminate arcing ground-faults and transient overvoltage

(h) To provide an equal voltage reference to all phase conductors to balance phase voltages

(i) To reduce harmonics at all frequencies with respect to ground for clean reliable control operation

Shipboard Power Systems Design and Verification Fundamentals, First Edition. Mohammed M. Islam.
© 2018 the Institute of Electrical and Electronics Engineers, Inc. Published 2018 by John Wiley & Sons, Inc.

(j) To provide energy absorption and protection for equipment within the CBEMA curve

(k) To provide an early warning signal and ground indication for arc-fault hazard warnings

The shipboard ungrounded power generation and distribution system-related ground detection and management systems considered are as follows:

(i) LV ground detection system with lights as outlined in IEEE 45.

(ii) Medium-voltage distribution ground detection system with high resistance grounding requirements, refer to IEEE 142.

(iii) Insulation monitoring system along with ground detection system, refer to IEC-61577.

(iv) Voltage-stabilizing ground reference system: commonly used for the ground fault detection of an ungrounded system

9.1 SYSTEM GROUNDING PER IEEE 45

9.1.1 Shipboard LV Power System Grounding IEEE 45 Recommendations (See Figures 9.1 and 9.2)

IEEE Std 45-2002: Ground detection for each ungrounded system should have a monitoring and display system that has a lamp for each phase that is connected between the phase and the ground. This lamp should operate at more than 5 W and less than 24 W when at one-half voltage in the absence of a ground. The monitoring and display system should also have a normally closed, spring return-to-normal switch between the lamps and the ground connection. If lamps and continuous ground monitoring utilizing superimposed DC voltage are installed, the test switch should give priority to continuous ground monitoring and should be utilized only to determine which phase has the ground fault by switching the lamp in. With continuous ground monitoring tied into the alarm and monitoring system, consideration will be given to alternate individual indication of phase to ground fault. If lamps with low impedance are utilized, the continuous ground monitoring is reading ground fault equivalent to the impedance of the lamps, which are directly connected to the ground.

Where continuous ground monitoring systems are utilized on systems where nonlinear loads (e.g., adjustable speed drives) are present, the ground monitoring system must be able to function properly.

IEEEStd 45-2002 CLAUSE-5.9.7.2 Ground detection lamps on ungrounded systems

Ground detection for each ungrounded system should have a monitoring and display system that has a lamp for each phase that is connected between the phase and the ground. This lamp should operate at more than 5 W and less than 24 W when at one-half voltage in the absence of a ground. The monitoring and display system should also have a normally closed, spring return-to-normal switch between the lamps and the ground connection.

If lamps and continuous ground monitoring utilizing superimposed dc voltage are installed, the test switch should give priority to continuous ground monitoring and should be utilized only to determine which phase has the ground fault by switching the lamps in.

With continuous ground monitoring tied into the alarm and monitoring system, consideration will be given to alternative individual indication of phase to ground fault. If lamps with low impedance are utilized, the continuous ground monitoring is reading ground fault equivalent to the impedance of the lamps, which are directly connected to the ground.

Where continuous ground monitoring systems are utilized on systems where non-linear loads (e.g., adjustable speed drives) are present, the ground monitoring system must be able to function properly.

Ungrounded 480V, 3 Phase System

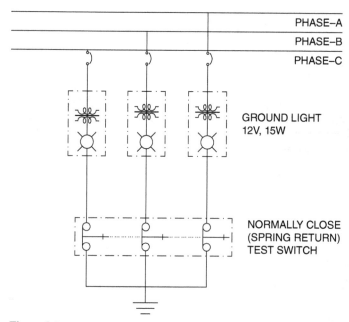

Figure 9.1   Typical Shipboard Ground Detection System for Ungrounded Generation and Distribution (Type-1).

Generator (Delta Connection–Ungrounded)
480V, 3 Phase System Shipboard Application

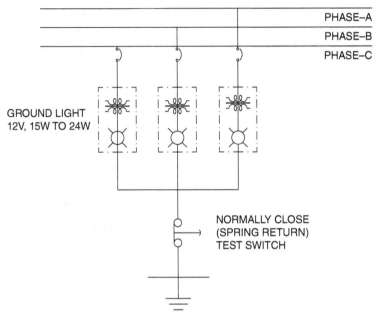

Figure 9.2 Typical Shipboard Ground Detection System for Ungrounded Generation and Distribution (Type-2).

9.2 SELECTION OF HIGH-RESISTANCE GROUNDING (HRG) SYSTEM

An ungrounded shipboard power system is used because under single-phase fault the system is considered operational for a short period. The single-phase fault is limited by the capacitance of the healthy phases. However, it must be noted that during the single-phase ground fault the voltage to ground of the ungrounded phases increases by the factor of $\sqrt{3}$ (see Fig. 9.3).

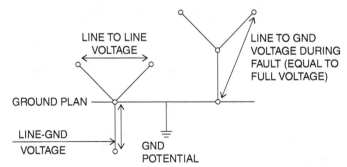

Figure 9.3 Typical Ungrounded Distribution Vector Diagram with Single-Phase Grounded (Type-1).

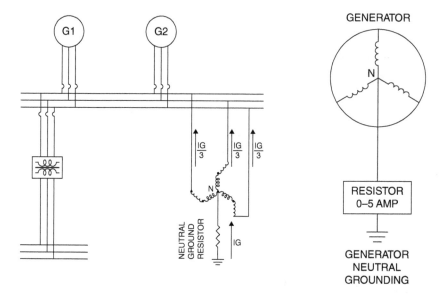

2. DISTRIBUTION BUS WITH ZIG-ZAG TRANSFORMER
   AND HIGH RESISTANCE GROUND

1. GENERATOR NEUTRAL
   GROUNDING RESISTOR

Figure 9.4   Typical Ungrounded System Generator Neutral Grounding and Bus Zigzag Transformer Grounding.

For a typical 6600 V system the line-to-ground (LG) voltage can be = 3810 V. In case of a single-phase fault, the line to ground of the other two phases will increase to $\sqrt{3} \times 3810 \text{ V} = 6600 \text{ V}$. This overvoltage may lead to insulation stress for all electrical equipment within that system. If the insulation rating is not sufficient to withstand the overstress voltage, the equipment will get overheated and ultimately can fail. If a second ground fault occurs before the first fault has been cleared, it can lead to an arcing fault. Therefore, for shipboard ground detection systems, the single-phase fault and phase-to-phase fault must be detected as soon as possible and protective measures must be taken to avoid an arcing explosion and equipment malfunction onboard ship.

9.3 IEEE 142 GROUND DETECTION REQUIREMENTS

Extract from IEEE Std 142-2007, Recommended Practice for Ground Detection System

Clause 1.4.3.1 High-resistance grounding (See Figs. 9.4, 9.6, and 9.7)

High-resistance grounding employs a neutral resistor of high ohmic value. The value of the resistor is selected to limit the current Ir to a magnitude equal to or slightly greater than the total capacitance charging current, 3 Ico.

Typically, the ground-fault current is limited to 10 A or less, although some specialized systems at voltages in the 15 kV class may require higher ground-fault levels.

In general, the use of high-resistance grounding on systems where the line-to-ground fault exceeds 10 A should be avoided because of the potential damage caused by an arcing current larger than 10 A in a confined space (see Foster, Brown, and Pryor).

Several references are available that give typical system charging currents for major items in the electrical system (see Electrical Transmission and Distribution Reference Book; Baker). These will allow the value of the neutral resistor to be estimated in the project design stage. The actual system charging current may be measured prior to connection of the high-resistance grounding equipment following the manufacturer's recommended procedures.

9.4 IEC REQUIREMENTS: INSULATION MONITORING SYSTEM

IEC requirements for a ground detection system include a System-level insulation monitoring system that goes beyond the requirements for a ground detection system. The details of IEC-61577-8 are as follows:

4.1 Insulation monitoring devices shall be capable of monitoring the insulation monitoring of IT systems including symmetrical and asymmetrical components and to give a warning if the insulation resistance between the system and earth falls below predetermined level.

Note-1: The symmetrical insulation deterioration occurs when the insulation resistance of all conductors in the system to be monitored decreases (approximately similarly). An asymmetric insulation deterioration occurs when the insulation resistance of (one conductor for example) occurs substantially more then the other conductor.

Note-2: So called earth fault relays using a voltage asymmetry (voltage shift) in the presence of an earth fault as the only measurement criteria, are not insulation monitoring devices in the interpretation of IEC 61577.

Note-3: A combination of several measurement methods including asymmetric monitoring may become necessary for fulfilling the task of monitoring special conditions on the system.

4.2 The insulation monitoring system shall comprise a test device, or be provided with a means for connection of a test facility for detecting whether the insulation monitoring device is capable of fulfilling its functions. The system to be monitored shall not be directly earthed and the devices shall not be suffered damage. This test is not intended for checking the response value.

4.3 Contrary to IEC-61577-1, the PE connection of insulation monitoring device is a measuring connection and may be treated as functional earth connection (FE).

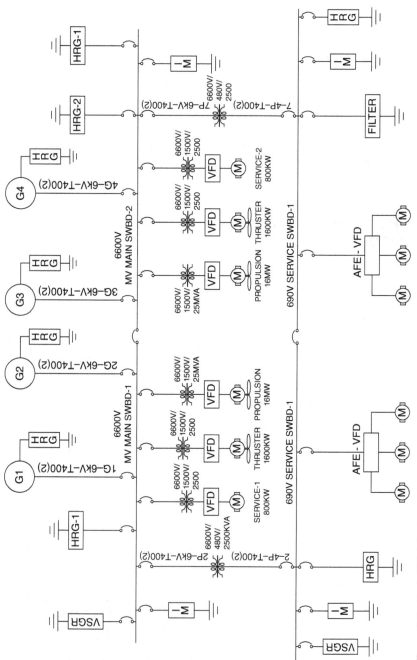

Figure 9.5 Typical High-Resistance Ground Detection System for Shipboard Power System.

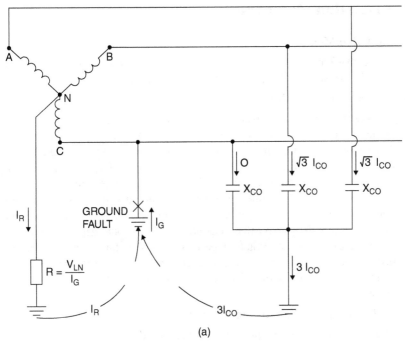

(a)

Figure 9.6   Single Line to Ground Fault on a Low-Resistance Grounding System for Marine Ungrounded System (Taken From IEEE-142 Figs. 1.6 and 1.9).

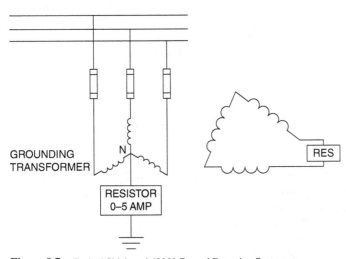

Figure 9.7   Typical Shipboard 480 V Ground Detection System.

9.4.1 Insulation Monitoring

(a) **Insulation Monitoring Device (IMD)**

The IMD is the device of choice for the protection of floating systems. IMDs come in two styles: A) Passive and B) Active devices.

(b) **Passive Insulation Monitoring Device (IMD)**

The best-known passive insulation monitoring device is the three lightbulb system in 480 V ungrounded systems. Three lamps are connected on their secondary side together and from there to ground (Star or Y configuration). Each lamp is then connected to the respective phases L1, L2, and L3. In a healthy system, all three lights will burn with the same intensity. In case of a ground fault, the faulted phase will assume a value close to ground potential. The respective light will dim, while the other two will brighten up. The lightbulb system often does not offer additional trip indicators for remote alarms. It also needs to experience a serious fault condition before people become aware that something is going wrong. Even worse, symmetrical ground faults (a balanced fault on all three phases) will not be detected.

(c) **Active Insulation Monitoring Device (IMD)**

The active insulation monitoring device is considered to be an online megger. It will be connected via pilot wires between the system and ground. A constant measuring signal will be sent from the IMD into the power wires. It will spread out evenly into the secondary side of the supply transformer and the attached loads. If this signal finds a break through path to ground, it will take this path of least resistance and return to the monitor. The IMD's internal circuitry will process the signal and trip a set of indicators when the set point is increased. IMDs measure in ohms (resistance) and not in amps (current). A ground fault will be indicated as "insulation breakdown."

9.4.2 Insulation Monitoring System for Grounded AC Systems with VFD System

The 60-cycle ground fault relay (GFR) has limitations when the circuitry involves VFDs (Variable Frequency Drives). Tests have shown that the typical GFR cannot keep the adjusted trip point when the system frequency changes to values below 60 cycles. Even worse, a total failure may occur at frequencies below 12 cycles. A variable frequency drive converts the incoming AC internally into DC, which is modulated again into a variable cycle AC leading to the load.

Some drives might be equipped with their own internal scheme to detect ground faults, which will eventually trip in the high ampere range. Early warning or personnel protection cannot be guaranteed in this case.

9.5 SYSTEM CAPACITANCE TO GROUND CHARGING CURRENT CALCULATION (TAKEN FROM IEEE 142 FIGS. 1.6 AND 1.9)

9.6 TOTAL SYSTEM CAPACITANCE CALCULATION

High-resistance grounding is recommended for systems where power interruption resulting from single line-to-ground fault tripping is detrimental to the process.

The maximum ground fault current allowed by the neutral grounding resistor must exceed the total capacitance to ground charging current of the system.

The total capacitance to ground charging current of a system can be measured or estimated.

Power should be off before making any connection before the test. All components and devices used should be rated properly for the system voltage.

One line will be bolted to ground using a fast-acting fuse rated for 10 A or less, a circuit breaker, a rheostat and an ammeter all in series.

The rheostat should be set at maximum resistance and the circuit breaker open.

The circuit breaker will then be closed and the rheostat resistance will be reduced slowly to zero. At this point the current circulating to the ground and indicated in the ammeter is the capacitance to ground charging current of the system.

To finish the test the rheostat will be returned to maximum resistance and the circuit breaker will be opened.

A quick estimate can be obtained by adding the charging current indicated in the following table according to the system kVA and the surge capacitors installed:

480 V systems

System kVA	0.5 A / 1000 kVA

2.4 kV systems

System kVA	0.75 A / 1000 kVA

4.16 kV systems

System kVA	1.0 A / 1000 kVA

For system voltages greater than 4.16 kV or when the estimated charging current is too close to 10 A a better estimate can be obtained by using the following tables according to the system voltage and the equipment installed.

480 V systems

Cables 2/0 to 3/0 MCM in trays	0.02 A / 1000 ft
Motors	0.01 A / 1000 HP

4.16 kV systems

Cable (shielded) #1 - 350 MCM	0.23 A / 1000 ft
Cable (non-shielded)	0.1 A / 1000 ft
Motors	0.05 A / 1000 HP

6.6 kV systems

Table 9.1   System-Level Charging Current Estimate

Cable 1000 MCM shielded	1.15 A / 1000 ft
Cable 750 MCM shielded	0.93 A / 1000 ft
Cable 350 MCM shielded	0.71 A / 1000 ft
Cable 4/0 AWG shielded	0.65 A / 1000 ft
Cable 2/0 AWG shielded	0.55 A / 1000 ft
Motors	0.15 A / 1000 HP

9.7 CALCULATE CAPACITIVE CHARGING CURRENT: (FOR A TYPICAL INSTALLATION)

High-resistance grounding refers to the grounding of the 6.6 kV system using three single-phase grounding transformers with primaries connected in star and neutral grounded through a suitable resistor.

The rating of the grounding transformer is selected to limit the ground fault current flow to a value equal to or slightly greater than the capacitive charging current of the 6.6 kV system. The total capacitive charging current of the 6.6 kV system is calculated as the sum of the capacitive charging currents of each individual 6.6 kV system component as follows.

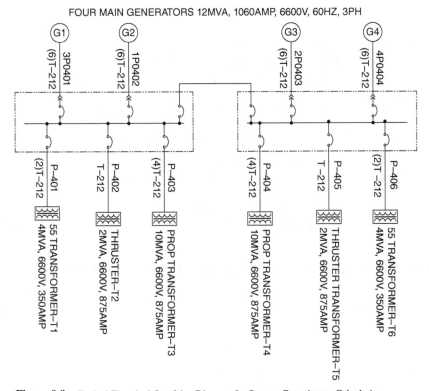

Figure 9.8   Typical Electrical One-Line Diagram for System Capacitance Calculation.

9.8 CAPACITIVE CHARGING CURRENT CALCULATION: SAMPLE CALCULATION

9.8.1 Iccc Calculation for Generators 12,000 kVA, 6600 V, 3-Phase, 3-Wire—Total 4

Capacitance per phase = 0.24 microFarad

Capacitive charging current – Ico (Gen1) = (2) * (π) * (f) * (c) * V/$\sqrt{3}$) = $2 \times \pi \times 60 \times 0.24 \times 10^{-6} \times 6600/\sqrt{3} = 0.326$ AMP

Total Capacitive charging current for 4 generators – I ccc (Gen) = $4 \times 0.326 = 1.304$ AMP/phase

9.8.2 Iccc Calculation for Transformers

(a) **10000 kVA Propulsion Transformers—Total 2 (T3-T4)**
Primary capacitance per phase = 4.0 nanoFarads
Capacitive charging current – Ico (PrT) = $2 \times \pi \times 60 \times 4.0 \times 10^{-9} \times 6600/\sqrt{3} = 5.746 \times 10^{-3}$ AMP

Total capacitive charging current for 4 off transformers – Iccc (PropT) = $2 \times 5.746 \times 10^{-3} = 0.0115$ AMP/phase

(b) 4000 kVA Ship Service Transformers—Total 2 off (T1 &T6)
Primary capacitance per phase = 4.3 nanoFarads
Capacitive charging current – Ico (SsT) = $2 \times \pi \times 60 \times 4.3 \times 10^{-9} \times 6600/\sqrt{3} = 6.177 \times 10^{-3}$ AMP
Total capacitive charging current for 4 off transformers - Iccc ((SST) = $2 \times 6.177 \times 10^{-3} = 0.0123$ AMP/phase

(c) 2000 kVA Thruster Transformers—Total 2 (T2 &T5)
Primary capacitance per phase = 4.3 nanoFarads
Capacitive charging current – Iccc (Thrus-T) = $2 \times \pi \times 60 \times 4.3 \times 10^{-9} \times 6600/\sqrt{3} = 6.177 \times 10^{-3}$ AMP
Total capacitive charging current for 4 off transformers – Iccc ((Thrus-T) = $2 \times 6.177 \times 10^{-3} = 0.0123$ AMP/phase

9.8.3 Cables 8 kV—(4/0 AWG) (T-212) Cable Three Core

Applicable Length:

Generator # 1	= 6 × 140 ft	= 840 ft
Generator # 2	= 6 × 120 ft	= 720 ft
Generator # 3	= 6 × 80 ft	= 480 ft
Generator # 4	= 6 × 80 ft	= 480 ft
Transformer – 1–3$'$	= 3 × 250 ft	= 750 ft
Transformer – 4–6	= 3 × 200 ft	= 600 ft
Transformer – 7–8	= 2 × 300 ft	= 600 ft
Harmonic Filter #1–2	= 2 × 50 ft	= 100 ft
Total Length of 212 MCM Cable		= 4570 ft

Capacitance per phase = 108 picoFarads/ft

Cable capacitive charging current = Ico = $2 \times \pi \times 60 \times 4570 \times 108 \times 10^{-12} \times 6600/\sqrt{3} = 0.481$ AMP/phase

9.8.4 Total Capacitive Charging Current

Total capacitive charging current is the summation of:

Generator – Icot (Gen1)	= 0.326 AMP/phase
Generator – Icot (Gen2)	= 0.326 AMP/phase
(2) Propulsion transformer – Icot (Prop -T)	= 0.023 AMP/phase
(2) Ship service transformer – Icot (SS-T)	= 0.0123 AMP/phase
(2) Thruster transformer – Icot (THR-T)	= 0.0123 AMP/phase
112.6 sq.mm 6.6 kV cable – Ico	= 0.481 AMP/phase
Therefore, total charging current	= 1.1806 AMP/phase

This is to set the resistor rating for the current to be 3 AMP or 5 AMP for the resistance grounding system

9.8.5 Grounding Transformer Size Calculation

(a) Low-Resistance Grounding: Transformer Sizing for 2400 V System

The neutral diving transformer for an ungrounded shipboard system is normally exposed to the full line-to-neutral voltage including any transients and momentary voltage excursions. Therefore, the protection of the HRG transformer should be taken under consideration.

For low-resistance grounding systems, at 2400 V, distribution of 300 A level, and transformer rating 2400/480 V (5:1 ratio), the secondary current will be $(100) \times (5) = 500A$, which will circulate in the broken delta windings. Each of three grounding transformer ratings will be $(100 \text{ A}) \times (2400 \text{ V}) = 240$ kVA. The transformer should be able to withstand the maximum duration of the ground fault, assuming a ground fault duration of 10 seconds.

To limit the ground transformer in a broken delta configuration for secondary current to 500 AMP, the resistor should be rated at $(480 \times \sqrt{3}) = 831$ V, and 500 AMP, $= 831/500 = 1.66$ ohms for 10 seconds at 831 voltage rating, knowing that the broken delta voltage during a ground fault would be 831 volts. The resistor voltage rating can be at the range of 600 V–1000 V.

When the broken delta voltage during a solid ground fault of 300 AMP will cause the neutral shift of $2400/\sqrt{3} = 1386$ V at the primary. For the ground fault of 150 AMP, the neutral shift will be half of $1386 = 693$ V at the primary and 415.5 at the open delta.

(b) High-Resistance Grounding: Transformer Sizing for 6600 V System

Based on the total capacitive charging current of 3 AMP/phase, the grounding transformer rating is $3 \times 6600 \approx 20$ kVA/phase (Note: The is transformer to be rated for line-to-line voltage in accordance with ANSI/IEEE Std C37.101 to withstand line-to-line voltage under ground fault conditions.)

Based on the total charging current of 3 amp/phase the single phase grounding transformer to be selected with the nearest tap set for the transformer line current of 3 AMP (i.e., for the ground current of $3 \times 3 = 9$ AMP) to provide the continuous rating of $6.6 \times 3 \approx 20$ kVA with a standard transformer ratio of 6600 V/120 V.

To ensure that the ground fault current is slightly greater than the 6.6 kV system capacitive charging current, the nearest standard transformer tap selected is for the line current of 3.3 AMP (ground current of $3 \times 3.3 \approx 10$ AMP), and this will give the transformer continuous rating of 22 kVA per phase.

9.8.6 Grounding Resistor Size Calculation

With all three single-phase transformers (each rating 22 kVA) connected in star on the primary side (6600 V) and neutral solidly grounded, and the secondary side connected in open delta, the secondary current is (based on the transformer primary line current of 3.3 AMP):

(6600/240) × 3.3 = 91.5 AMP

Maximum open delta voltage under ground fault is $\sqrt{3} \times 240 = 415.7$ V

Loading Resistor required = 415.7/91.5 = 4.5 ohms

Therefore, continuous rating of the resistor is = 415.7 × 91.5/1000 = 38 kW

Finally the presence of a resistance with a high value does not alter the isolation of the propulsion drive from the ground.

For a typical design the following can be considered:

R_{mas} = 233 ohms

$I_{continuous}$ = 3 AMP

I_{peak} = 5 AMP for 1 second

Insulation voltage = > 3,6 kV

9.9 GROUNDING RESISTOR SELECTION GUIDELINE PER IEEE STD 32-1972

IEEE 32 is the standard used for rating and testing neutral grounding resistors. The most important parameters to consider:

– allowable temperature rises of the element for different "on" times;
– applied potential tests; the dielectric tests, and
– resistance tolerance tests that are required.

- Time Rating: Neutral Grounding Resistor (NGR)
  IEEE Standard 32 specifies standing time ratings for Neutral Grounding Resistors (NGRs) with permissible temperature rises above 30 °C ambient. Time ratings indicate the time the grounding resistor can operate under fault conditions without exceeding the temperature rises.

- 10-Second Rating
  This rating is applied on NGRs that are used with a protective relay to prevent damage to both the NGR and the protected equipment. The relay must clear the fault within 10 seconds.

- One-Minute Rating
  One NGR is often used to limit ground current on several outgoing feeders. This reduces equipment damage, limits voltage rise and improves voltage regulation. Since simultaneous grounds could occur in rapid succession on different feeders, a 10-second rating is not satisfactory. The one-minute rating is applied.

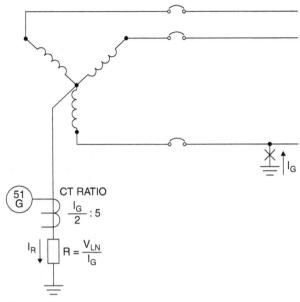

Figure 9.9 IEEE 142 (Figure 1.10 Scheme for Detecting a Ground Fault on a Low-Resistance Grounding System for a Marine Ungrounded System).

9.10 GROUNDING RESISTOR DUTY RATING

Short time rating: Short time ratings are 0–10 seconds or 0–60 seconds. Since short time rated resistors can only withstand rated current for short period of time, they are usually used with fault clearing relays. The short time temperature rise for the resistive element is 760C.

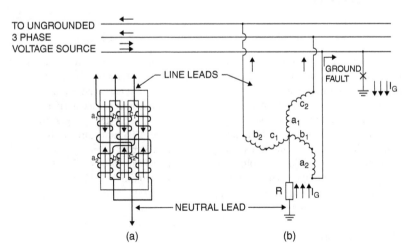

Figure 9.10 IEEE 142 (Figure 1.14(1) Zigzag grounding transformer: (a) core winding (b) system connection for Marine Ungrounded System).

6600V UNGROUNDED SYSTEM HIGH RESISTANCE GROUNDING
SYSTEM NORMAL OPERATION AND GROUND–FAULT –PHASE A

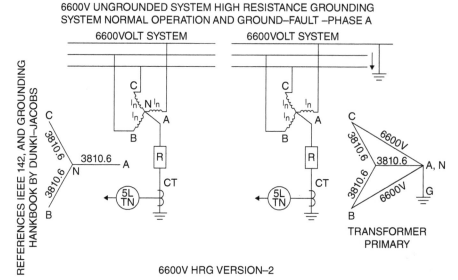

Figure 9.11   6.6 KV System with HRG Showing Voltage Vectors (Version-2).

Extended time rating: A time rating greater than 10 minutes that permits the temperature rise of resistive elements to become constant, but limited to an average not more than ninety days per year. The extended temperature rise for the resistive element is 610C.

Continuous rating: Capable of withstanding rated current for an indefinite period of time. The continuous temperature rise for the resistive element is 385C.

9.11 ZIGZAG GROUNDING TRANSFORMERS: IEEE STD 142 SECTION 1.5.2

One type of grounding transformer commonly used is a three-phase zigzag transformer with no secondary winding. The internal connection of the transformer is illustrated in Figure 1.14(1). The impedance of the transformer to balanced three-phase voltages is high so that when there is no fault on the system, only a small magnetizing current flows in the transformer winding. The transformer impedance to zero-sequence voltages, however, is low so that it allows high ground-fault currents to flow. The transformer divides the ground-fault current into three equal components; these currents are in phase with each other and flow in the three windings of the grounding transformer. The method of winding is seen from Figure 1.14(1) to be such that when these three equal currents flow, the current in one section of the winding of each leg of the core is in a direction opposite to that in the other section of the winding on that leg. This tends to force the ground-fault current to have equal division in the three lines and accounts for the low impedance of the transformer-to-ground currents.

A zigzag transformer may be used for effective grounding, or an impedance can be inserted between the derived neutral of the zigzag transformer and ground to obtain the desired method of grounding. This transformer is seldom employed for medium-voltage, high-resistance grounding. An example of low-resistance grounding is shown in Figure 1.14(2). The overcurrent relay, 51G, is used to sense neutral current that only flows during a line-to-ground fault.

9.12 RATING AND TESTING NEUTRAL GROUNDING RESISTORS: IEEE STD 32-1972

Rating and Testing Neutral Grounding Resistors: IEEE Std 32-1972:
    IEEE 32 is the standard used for rating and testing neutral grounding resistors. The most important parameters to consider from IEEE 32 are: the allowable temperature rises of the element for different "on" times; the applied potential tests; the dielectric tests, and the resistance tolerance tests that are required. Post Glover Neutral Grounding Resistors are designated and built to pass all these rigorous tests.

- Time Rating
  IEEE Standard 32 specifies standing time ratings for Neutral Grounding Resistors (NGRs) with permissible temperature rises above 30 °C.
      Time ratings indicate the time the grounding resistor can operate under fault conditions without exceeding the temperature rises.

- 10-Second Rating
  This rating is applied on NGRs that are used with a protective relay to prevent damage to both the NGR and the protected equipment. The relay must clear the fault within 10 seconds.

- One-Minute Rating
  One NGR is often used to limit ground current on several outgoing feeders. This reduces equipment damage, limits voltage rise, and improves voltage regulation. Since simultaneous grounds could occur in rapid succession on different feeders, a 10-second rating is not satisfactory. The one-minute rating is applied.

- Ten-Minute Rating
  This rating is used infrequently. Some engineers specify a 10-minute rating to provide an added margin of safety. There is, however, a corresponding increase in cost.

- Extended-Time Rating
  This is applied where a ground fault is permitted to persist for longer than 10 minutes, and where the NGR will not operate at its temperature rise for more than an average of 90 days per year.

- Steady-State Rating
  This rating applies where the NGR is expected to be operating under ground fault conditions for more than an average of 90 days per year and/ or it is desirable to keep the temperature rise below 385 °C.

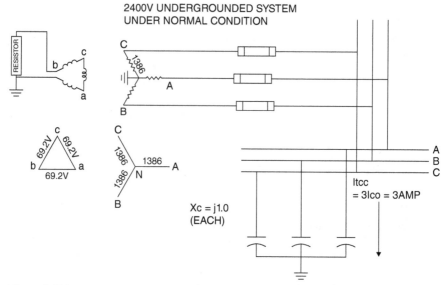

Figure 9.12 2400 V System with HRG Ground Detection: Normal Operational Condition.

Tests

An applied potential test (HI-POT) is required to test the insulation of the complete assembly (or sections thereof).

For 600 volts or less, the applied potential test is equal to twice the rated voltage of the assembly (or section) plus 1,000 volts.

For ratings above 600 volts, the applied potential test is equal to 2.25 times the rated voltage, plus 2,000 volts.

The resistance tolerance test allows plus or minus 10 percent of the rated resistance value.

9.13 VOLTAGE STABILIZING GROUND REFERENCE (VSGR) PHASEBACK FOR GROUND DETECTION (CURTSEY OF APPLIED ENERGY)

Phaseback is a patented product created by Applied Energy LLC. Phaseback is used for voltage stabilizing ground reference (VSGR) by monitoring the voltage on each phase with respect to ground and electromagnetically reacting to any voltage imbalance in the system and then stabilizing the ground reference of the system.

Ungrounded shipboard power systems require special consideration because there is no safe way to ground the system without inducing current and violating the National Electric Code (NEC), which states *"ground is not to be an intentional current carrying conductor."* It is dangerous to allow current to flow to ground when intermittent faults, transient overvoltage occurrences, and short-circuit problems can create potentially dangerous conditions for personnel.

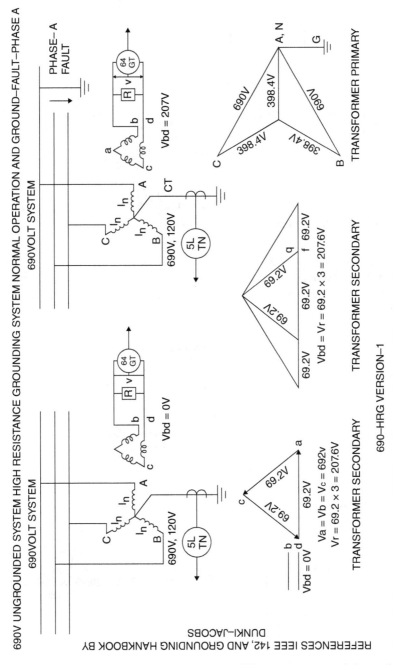

Figure 9.13 690 V System with HRG Ground Detection: Normal and Faulted Conditions.

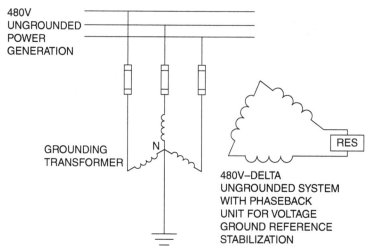

Figure 9.14 480 V System with Voltage Stabilizing Unit.

It is necessary to provide a ground reference for control systems to properly operate while preventing current to flow to ground, which could cause safety concerns, and then a second phase connecting to ground can cause a possible arc flash.

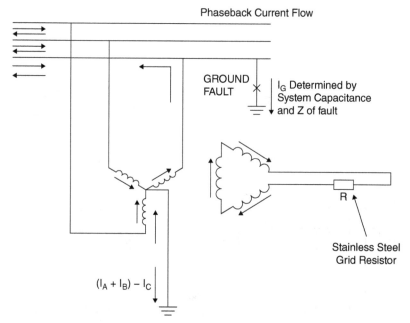

Figure 9.15 Voltage Stabilizing Ground Refernce: Phaseback Shown One Phase Grounded.

The following is a short explanation to understand the operation of the Phaseback voltage stabilizing ground reference to establish and maintain a stable equal voltage from each phase to ground:

Ohm's Law for AC circuits is $(V = IZ)$, Voltage equals Current multiplied by Ohms of Impedance. As the leading event, the voltage change, when detected, offers the fastest reaction time.

The Voltage Stabilizing Ground Reference (VSGR) Phaseback unit operates very similar to a braking resistor, dissipating the charge current in the power system through the resistor in the series secondary circuit to lower the capacitive energy to a safe level, thus mitigating the arc flash potential during a power shut down.

The Voltage Stabilizing Ground Reference (VSGR) unit Phaseback makes a three-phase wye connected transformer from three single-phase transformers with the center-tap or neutral of a wye connected primary is connected to ground. The ungrounded single-phase secondary coils are connected in series with a large power resistor connected in series with those secondary coils without any ground connection.

The secondary series circuit in Phaseback has the same current through the entire circuit. Circuits with equal ohms of impedance and equal current will maintain equal voltage. The transformer couples the primary and secondary circuits with magnetic lines of force. The secondary current of each transformer controls the primary current of that transformer and the voltage ratio is equal to the turns ratio in the transformer. The current ratio is at the inverse of the voltage ratio in the transformer.

The large power resistor limits current to protect the transformers from overload. The current through and voltage across the resistor are in-phase (100% power factor).

Harmonics are reduced and filtered through the resistor in the transformer secondary as current circulates.

The transformers are identical with equal secondary voltages and equal primary voltages. The transformers are bidirectional: step-down and step-up and the phase voltages across each primary coil are equal when the power system is balanced and cause the secondary voltages in the Voltage Stabilizing Ground Reference (VSGR) to be equal.

During an imbalance in the power distribution system, current flow in the secondary back to the low-voltage primary phase and the positive feedback keeps the capacitive charge equal, causing the phase voltages on the power system to be stable.

The Phaseback–Voltage Stabilizing Ground Reference (VSGR) unit provides an alarm on the front panel with a ground fault indicating light and a circuit current detector with solid-state (normally open) contacts. The current detector is factory set to alarm when 500 milliamps flow through the resistor. The early warning of the current detector provides the ability to correct the ground problem before a second fault forms, thus preventing the cause of arc flash.

In the event the ground cable is cut or otherwise disconnected, backup protection can be provided by connecting a second current detector on the ground connection that is connected to the neutral point of the primary connection in the unit.

9.13.1 Typical HRG Elementary Diagrams Are Very Close to the Voltage Stabilizing Ground Reference (VSGR) Phaseback Unit, Looking Alike Phaseback's Function Is Exactly Opposite to the HRG

a. Typical HRG units are designed to control and monitor ground current. Phaseback does not monitor ground current; phaseback monitors voltage and tries to balances phase voltage during a phase ground.

b. HRG units typically use wire wound resistors, which add inductance (inductors are coiled conductors) into the circuit. Phaseback uses noninductive resistors. This allows Phaseback to push current back into the low-voltage phase at the correct time to balance phase voltages. When HRG inductance is entered into the circuit, it can cause the current to lag the voltage and it will not balance the voltages when a ground fault deepens.

c. HRG units have a ground on the secondary of the three single-phase transformers, even when this is not shown on the elementary diagram. This locks the center (neutral) of the primary and one corner of the secondary in phase as they share a ground point. The transformer primary and secondary are no longer isolated as they are coupled through the common ground point. The noise the secondary is filtering is injected back into the primary circuit. This also prevents the HRG unit from properly balancing the voltage as a ground fault deepens.

d. There are similarities in the appearance of the circuits on paper, but there is a big difference in the way they function. They both can indicate the presence of a ground fault; Phaseback corrects the problem by balancing the voltages with respect to ground to prevent both voltage spikes and arcing ground faults.

e. In the described operation of HRG units, the ground current from the three single-phase primaries are added equally. This is the ground current monitored with the current transformer and displayed on the ammeter. The ground wire in Phaseback does not have the sum of the currents: it has the difference. Typically, the ground current using Phaseback is from one-third to less than one-half the current of the HRG.

f. Less ground current means less hull current, which means less electrical noise.

g. Voltage Stabilizing Ground Reference (VSGR)–Phaseback has proven to reduce or eliminate the arc -flash hazard and minimize the danger to people during arcing ground -faults, preventing them from being solid grounds.

h. Voltage Stabilizing Ground Reference (VSGR)–Phaseback removes approximately 85% of the phase voltage harmonics by keeping the phase voltages balanced.

i. Voltage Stabilizing Ground Reference (VSGR)–Phaseback prevents voltage spikes from forming, and makes power systems immune to voltage spikes from external sources.

j. When a Voltage Stabilizing Ground Reference (VSGR)–Phaseback is installed to an ungrounded power system, there is no longer a need for other power filtering or protection devices, as each Phaseback has all of these functions and features built into it: (a) Additional ground indicators, (b) Insulation monitors, (c) TVSS units, (d) Harmonic mitigation devices or systems, (e) Additional shunt trip devices or controls.

9.13.2 Phaseback Voltage Stabilizing Ground Reference Addresses and Solves the Following Issues

(a) Arc flash prevention and mitigation

(b) Voltage spikes from internal or external sources

(c) Phase voltage imbalance

(d) Phase loss due to high impedance grounds

(e) Phase angle differential

(f) Phase voltage instability

(g) Phase voltage harmonics

(h) Waveform distortion

(i) Noisy ground reference and frequency instability

(j) Arcing ground faults

(k) Operational efficiency increases

HRG-Neutral Grounding Transformer Example—6.6 kV System (Typical)

Rated system (primary) voltage 6600 V and secondary voltage 120 V

Connections star—and open delta

Rated current—5 AMP

Rated time-continuous (10 seconds, or 60 seconds)

Number of phases—3 phases

Rated frequency—60 Hz

Insulation class—F

Standards–IEC 60726

HRG-Neutral Grounding Resistor—120 V (Typical)

Voltage—120 V

Rated time—continuous

Continuous current—4.05 AMP (10 seconds, or 60 seconds)

Resistance at 25 °C—381 ohms (+ −10%)

Maximum element temp rise—280 °C

Resistance change for 200 °C rise

9.14 HRG VERSUS VSGR

A simplified shipboard electrical power system topology is shown to provide better understanding of various types of shipboard groundings.

Table 9.2 Methods of Different Grounding Systems and their Characteristics (Grounding Options as Applicable for Specific Installation Requirements)

System Characteristics	Ungrounded System with HRG	Ungrounded System with VSGR
Continuity of service on ground fault	*Yes*	*Yes*
Capability to propagate multiphase fault	Low	Lowest
Equipment damage potential	Very Low	Lowest
Ground fault current	Low	Lowest
Transient overvoltage level	$\geq 6\,X$	1 X
Cable insulation level	1.73	1.0
Arc flash risk level	Very Low	Lowest

9.15 SHIPBOARD GROUND DETECTION SYSTEM RECOMMENDATIONS

Due to the fact that shipboard unground system ground detection is very complex, there can be no universal solution for the ground detection system. Therefore, the shipboard ground detection system development for specific applications must be developed by the experts and approved by the authority having jurisdiction. Above all, the ground detection system must be approved by regulatory bodies in view of the safety of the system and the safety of the operators. There must be appropriate alarms and monitoring to ensure early detection, monitoring, and if necessary the shutdown of the grounded system.

REFERENCES

[1] IEEE 32: Neutral Grounding Resistors
[2] IEEE 45
[3] IEEE 142
[4] IEEE 1580
[5] IEC-60726
[6] Westinghouse, "System Neutral Grounding and Ground Fault Protection," publication PRSC-4B-1979, Westinghouse, 1979.

[7] General Electric Co., *"Generator Neutral Grounding,"* publication GET-1941, Schenectady, NY: General Electric.

[8] "Charging Current Data for Guesswork-Free Design of High-Resistance Grounded Systems," by D.S. Baker, *IEEE Transactions on Industry Applications*, Vol. IA-15, No. 2, March/April 1979.

[9] Baldwin Bridger, Jr., High-Resistance Grounding, *IEEE Transactions on Industry Applications*, Vol. IA-19, No. 1, Jan/Feb 1983.

Chapter 10

Shore Power LV and MV Systems

10.0 INTRODUCTION

Shipboard shore power is 480 V, three-phase, 60 Hz, ungrounded. This is to correspond to the 450 V power generation and distribution on shipboard. Similarly 230 V shore power is introduced for ship with 230 V power generation and distribution.

Due to high power requirements onboard ships, medium-voltage and high-voltage shore power have been introduced. The medium-voltage shore power standard IEC/ISO/IEEE 80005.1 was developed jointly by the IEC, ISO, and IEEE.

10.1 LV SHORE POWER SYSTEM

Typical LV shore power electrical one-line diagrams are shown in the following figures:

Figure 10.1: 230 V ungrounded system with two shore power

Figure 10.2: typical 450 V ungrounded system with one shore power

Figure 10.3: typical ungrounded LV system shore power connection details with instrumentation

Figure 10.4: typical ungrounded 208 V system shore power connection details

Figure 10.5: typical 4 wire grounded LV system shore power connections

Figure 10.6: 480 V one shore with cable identification

Figure 10.7: 480 V three shore with three dedicated circuit breakers and separate feeders

Figure 10.8: 480 V three shore with one dedicated circuit breaker and three separate feeders

Figure 10.9: 480 V shore with one dedicated circuit breaker and four separate feeders for four receptacles

Shipboard Power Systems Design and Verification Fundamentals, First Edition. Mohammed M. Islam.
© 2018 the Institute of Electrical and Electronics Engineers, Inc. Published 2018 by John Wiley & Sons, Inc.

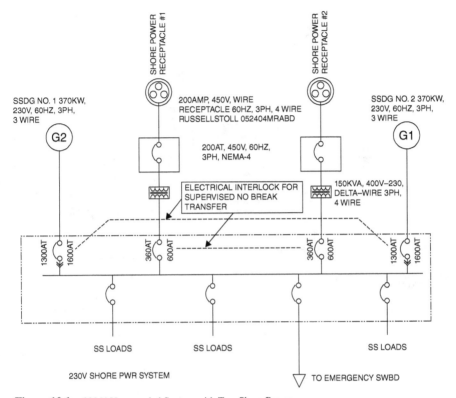

Figure 10.1 230 V Ungrounded System with Two Shore Power.

10.2 MV (HV) SHORE POWER SYSTEM

Due to recent high power requirements onboard ship, the shore power requirements also have changed. The IEEE 1713 standard development project was authorized by IEEE for medium-voltage shore power requirements. However, IEEE 1713 was replaced with 80005.1 as a decision was made to publish a joint IEC, ISO, and IEEE standard.

IEC/ISO/IEEE 80005-1 Edition 2012 International Standard: Utility connections in port—Part 1: High-Voltage Shore Connection (HVSC) Systems—General requirements:

Doc 80005-1 Figure 1: Block diagram of a typical described HVSC system arrangement

Annex B: Additional requirements for roll-on roll-off (RORO) cargo ships and RORO passenger ships

Annex C: Additional requirements for cruise ships

Figure C.2: Cruise ship HVSC system single-line diagram

Figure C.3: Example of safety and control circuit

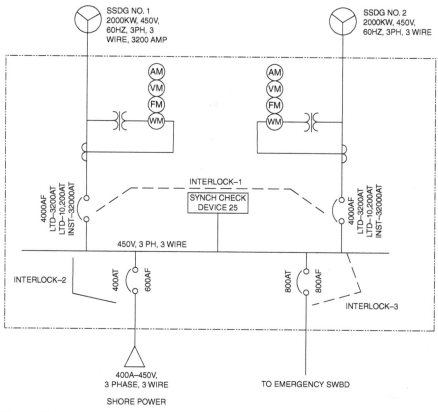

Figure 10.2 Typical 450 V Ungrounded System with One Shore Power.

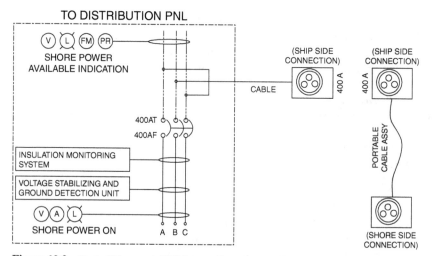

Figure 10.3 Typical Ungrounded LV System Shore Power Connection Details with Instrumentation.

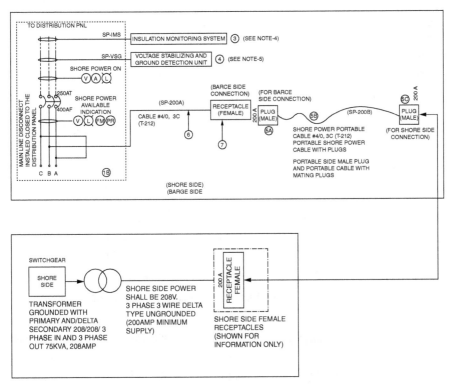

Figure 10.4   Typical Ungrounded 208 V System Shore Power Connection Details.

Annex D: Additional requirements of container ships

Annex E: Additional requirements of liquefied natural gas carriers (LNGC)

Typical MV (HV) shore power electrical one-line diagrams are shown in the following figures:

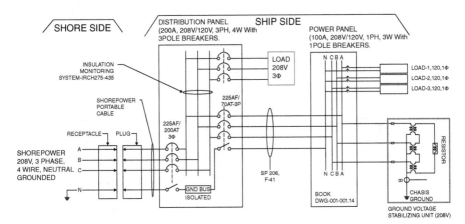

Figure 10.5   Typical 4-Wire Grounded LV System Shore Power Connections.

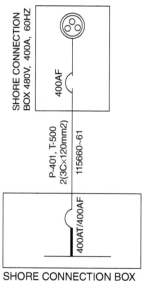

SHORE CONNECTION BOX
480V, 400A, 60HZ

Figure 10.6 480 V One Shore with Cable Identification.

Figure 10.10: Medium-voltage shore power with 300AMP plugs and receptacles type-1

Figure 10.11: Medium-voltage shore power with 300AMP plugs and receptacles type-2

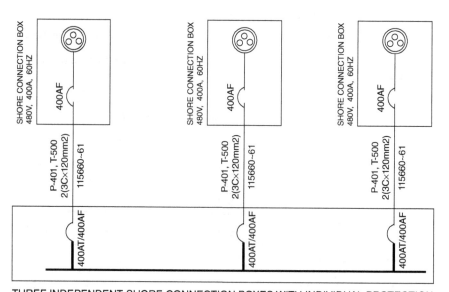

THREE INDEPENDENT SHORE CONNECTION BOXES WITH INDIVIDUAL PROTECTION

Figure 10.7 480 V Three Shore with Three Dedicated Circuit Breakers and Separate Feeders.

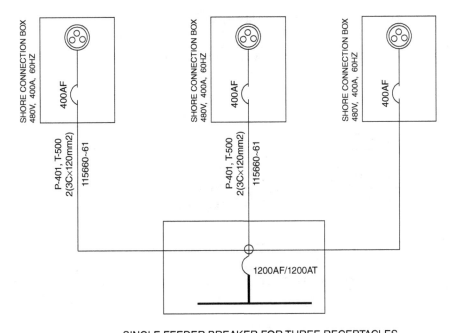

SINGLE FEEDER BREAKER FOR THREE RECEPTACLES

Figure 10.8 480 V Three Shore with One Dedicated Circuit Breaker and Three Separate Feeders.

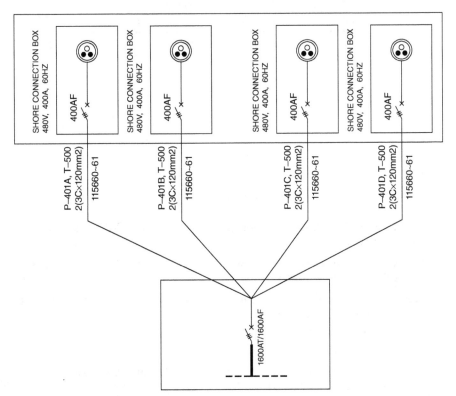

Figure 10.9 480 V Shore with One Dedicated Circuit Breaker and Four Separate Feeders for Four Receptacles.

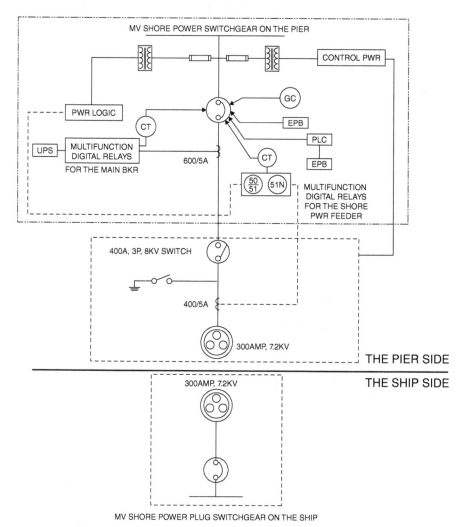

Figure 10.10   Medium-Voltage Shore Power with 300 AMP Plugs and Receptacles Type-1.

Figure 10.12: Medium-voltage shore power with two 300 AMP plugs and receptacles type-3

Figure 10.13: Typical MV shore power for cargo ship and RORO passenger ship per IEC/ISO/IEEE-80005-1

Figure 10.14: Typical MV shore power for cargo ship and RORO passenger ship per IEC/ISO/IEEE-80005-1

Figure 10.15: Typical MV shore power for cargo ship and RORO passenger ship per IEC/ISO/IEEE-80005-1

Figure 10.16: Typical MV shore power for cargo ship and RORO passenger ship per IEC/ISO/IEEE-80005-1

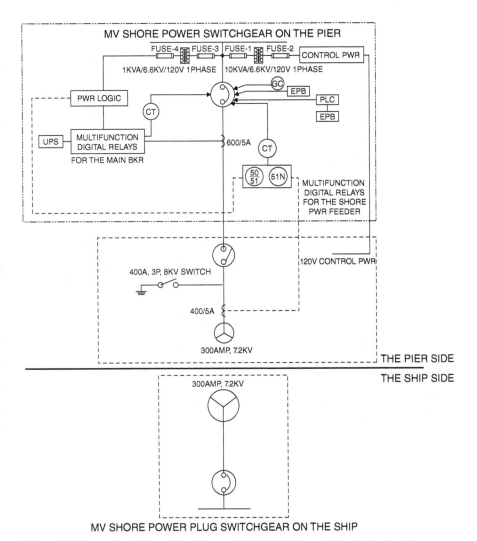

Figure 10.11   Medium-Voltage Shore Power with 300 AMP Plugs and Receptacles Type-2.

Figure 10.17: Typical MV shore power for cargo ship and RORO passenger ship per IEC/ISO/IEEE-80005-1

Figure 10.18: Typical MV shore power for cargo ship and RORO passenger ship per IEC/ISO/IEEE-80005-1

Figure 10.19: Medium-voltage shore power schematics for RORO cargo and passenger ships per IEC/ISO/IEEE-80005-1

Figure 10.20: Medium-voltage shore power schematics for container ship per IEC/ISO/IEEE-80005-1

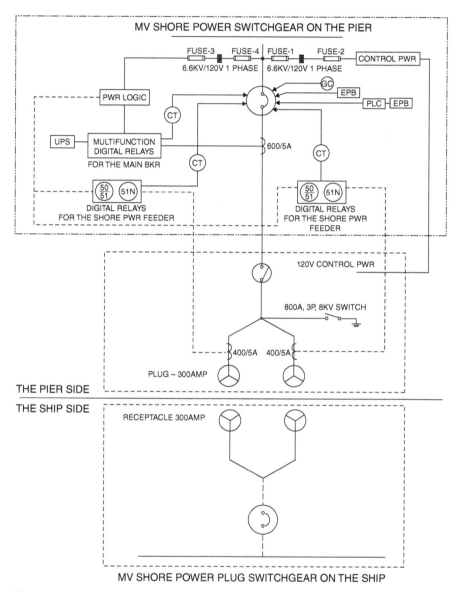

Figure 10.12   Medium-Voltage Shore Power with Two 300 AMP Plugs and Receptacles Type-3.

10.3 LOW-VOLTAGE SHORE POWER SYSTEM

The low voltage shore power is 230 V, three-phase ungrounded or 480 V, three-phase ungrounded. The 480 V shore power standard plugs and receptacles are 400 AMP with THOF-500 flexible cable. There are multiple 400 AMP plugs and receptacles to meet shipboard power demand. Figures 10-1 through 10-9 are typical low-voltage shore power installations.

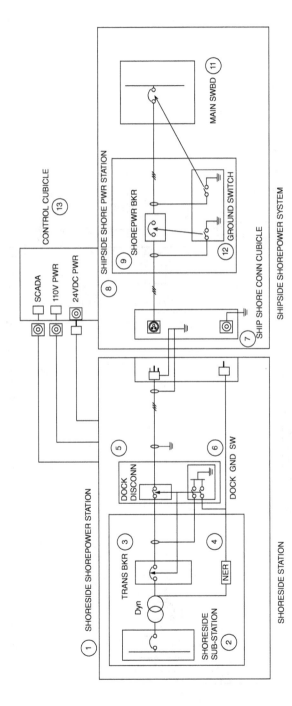

Figure 10.13 Typical MV Shore Power for Cargo Ship and RORO Passenger Ship per IEC/ISO/IEEE-80005-1.

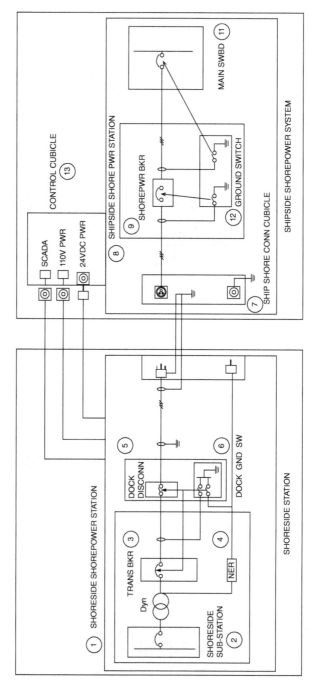

Figure 10.14 Typical MV Shore Power for Cargo Ship and RORO Passenger Ship per IEC/ISO/IEEE-80005-1.

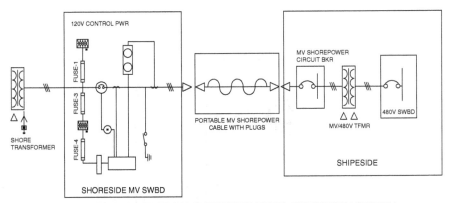

SHOREPOWER FROM SHORESIDE MV TO SHIPSIDE LV SYSTEM

Figure 10.15   Typical MV Shore Power for Cargo Ship and RORO Passenger Ship Per IEC/ISO/IEEE-80005-1.

10.4 FOUR-WIRE GROUNDED SYSTEM LV SHORE POWER CONNECTIONS

10.5 MEDIUM-VOLTAGE SHORE POWER SYSTEM (MV)

Medium-voltage shore power systems are more complex than the low-voltage system. This section is dedicated to explain various medium-voltage shore power recommendations as outlined in IEEE/IEC Standard 80005-1. Figures 10.10 through 10.20 are mostly extracted from the IEC/ISO/IEEE 8005.1 standard.

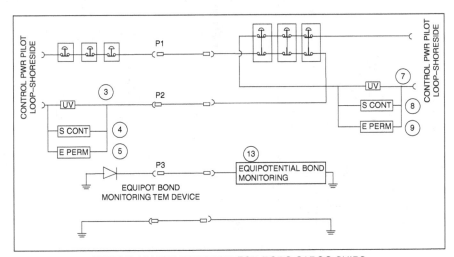

TYPICAL MV SHORE POWER FOR RORO CARGO SHIPS
AND RORO PASSANGER SHIPS PER IEC–IEEE–80005–1

Figure 10.16   Typical MV Shore Power for Cargo Ship and RORO Passenger Ship per IEC/ISO/IEEE-80005-1.

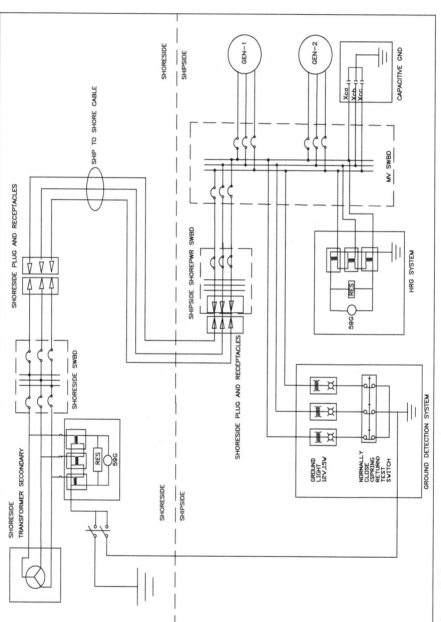

Figure 10.17 Typical MV Shore Power for Cargo Ship and RORO Passenger Ship per IEC/ISO/IEEE-80005-1.

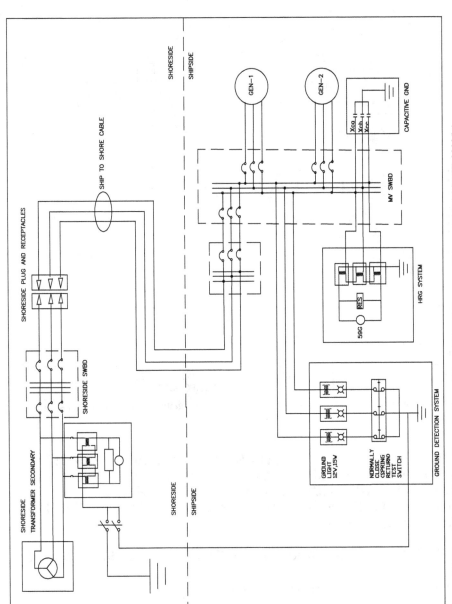

Figure 10.18 Typical MV Shore Power for Cargo Ship and RORO Passenger Ship per IEC/ISO/IEEE-80005-1.

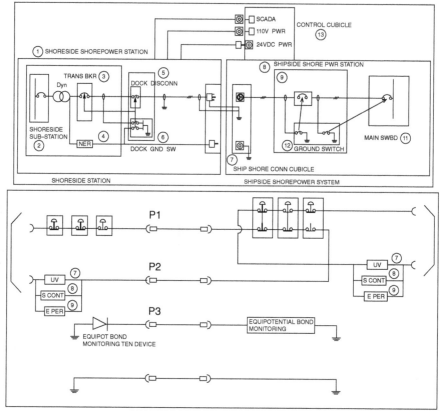

TYPICAL MV SHORE POWER FOR RORO CARGO SHIPS
AND RORO PASSANGER SHIPS PER IEC–IEEE–80005–1

Figure 10.19   Medium-Voltage Shore Power Schematics for RORO Cargo and Passenger Ships per IEC/ISO/IEEE-80005-1.

10.6 EXTRACT FROM IEC/ISO/IEEE 80005-1 PART 1: HIGH-VOLTAGE SHORE CONNECTION (HVSC) SYSTEMS HV SHORE POWER REQUIREMENTS (SHORE TO SHIP POWER QUALITY AND PROTECTION REQUIREMENTS)

IEEE-IEC 80005-1 Section 5.2 Quality of HV Shore Supply

The HV shore supply system shall have a documented voltage supply quality specification.

Ship electrical equipment shall only be connected to shore supplies that will be able to maintain the distribution system voltage, frequency and total harmonic distortion characteristics given below. For compliance, the compatibility assessment referred to in 4.3 shall include verification of the following:

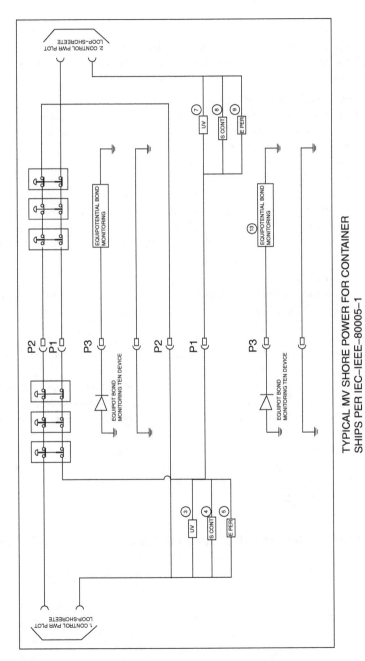

TYPICAL MV SHORE POWER FOR CONTAINER SHIPS PER IEC–IEEE-80005–1

Figure 10.20 Medium-Voltage Shore Power Schematics for Container Ship per IEC/ISO/IEEE-80005-1.

(a) *voltage and frequency tolerances (continuous):*

(1) *the frequency shall not exceed the continuous tolerances 5% between no-load and nominal rating;*

(2) *for no-load conditions, the voltage at the point of the shore supply connection shall not exceed a voltage increase of 6% of nominal voltage;*

(3) *for rated load conditions, the voltage at the point of the shore supply connection shall not exceed a voltage drop of −3,5% of nominal voltage.*

(b) *voltage and frequency transients:*

(1) *the response of the voltage and frequency at the shore connection when subjected to an appropriate range of step changes in load shall be defined and documented for each HV shore supply installation;*

(2) *the maximum step change in load expected when connected to a HV shore supply shall be defined and documented for each ship. The part of the system subjected to the largest voltage dip or peak in the event of the maximum step load being connected or disconnected shall be identified;*

(3) *comparison of (1) and (2) shall be done to verify that the voltage transients limits of voltage + 20% and % and the frequency transients limits of 10%, will not be exceeded.*

(c) *harmonic distortion:*

– *for no-load conditions, voltage harmonic distortion limits shall not exceed 3% for individual harmonic and 5% for total harmonic distortion.*

The HV shore supply shall include appropriate rated surge arrestors to protect against fast transient overvoltage surges (e.g., spikes caused by lightning strikes or switching surges).

Different voltage and frequency tolerances may be imposed by the owners or authorities responsible for the shore supply system and these should be considered as part of the compatibility assessment to verify the effect on the connected ship load is acceptable.

Where the possible loading conditions of a ship when connected to a HV shore supply would result in a quality of the supply different from that specified in IEC 60092-101:2002, 2.8, due regard should be given to the effect this may have on the performance of equipment.

IEC 800050-1 Section 4.8 System Study and Calculations

The shore-connected electrical system shall be evaluated. The system study and calculations shall determine:

(a) *the electrical load during shore connection;*

(b) *the short-circuit current calculations (see IEC 61363-1) shall be performed that take into account the prospective contribution of the shore supply and the ship installations. The following ratings shall be defined and used in these calculations:*

(1) *for shore supply installations, a maximum and minimum prospective short circuit current for visiting ships;*

(2) *for ships, a maximum and minimum prospective short circuit current for visited shore supply installations.*

(c) *the calculations may take into account any arrangements that:*

(1) *prevent parallel connection of HV shore supplies with ship sources of electrical power; and/or*

(2) *restrict the number of ship generators operating during parallel connection to transfer load;*

(3) *restrict load to be connected.*

(d) *system charging (capacitive) current for shore and ship;*

(e) *this system charging current calculation shall consider the shore power system and the expected ship power including the on line generator(s);*

(f) *shore power transformer neutral grounding resistor analysis; and*

(g) *transient overvoltage protection analysis.*

Chapter 11

Smart Ship System Design (S3D) and Verification

11.0 INTRODUCTION

Shipboard propulsion and auxiliary systems are increasingly reliant on electrical systems and electronic systems. Reliance on electronic technology from a distributed control system, to PLC, and then a microprocessor-based system has dramatically increased the complexity of the overall shipboard electrical system.

Smart Ship System Design (S3D) is a shipboard electrical system design environment with physics-based simulation and virtual prototyping of overall ship design, which is then compared with the real system of system for electrical power generation, distribution, protection, and automation.

S3D starts from the concept level of the design and continues through the preliminary design, and the detail design.

S3D provides ship system physics-based modeling and simulation prototyping for the following as a minimum:

- Electric plant dynamic load analysis
- Electric plant dynamic load flow calculation
- Electric plant short-circuit analysis
- Electric protection devices coordination and safety optimization
- Arc flash analysis and safety optimization
- Black start recovery with multiple scenario
- Harmonic analysis with system-level effect to minimize electrical noise
- Generator cold starting (black start)
- Frequency-dependent machine models
- Frequency-dependent network models

Shipboard Power Systems Design and Verification Fundamentals, First Edition. Mohammed M. Islam.
© 2018 the Institute of Electrical and Electronics Engineers, Inc. Published 2018 by John Wiley & Sons, Inc.

– Determination of optimal electrical loading time

– Generator and motor starting behavior

– Governor and automatic voltage regulator (AVR) behavior

– Power recovery to critical loads after loss of power

– System-level grounding requirement analysis

Recent electrical failure-related incidents onboard passenger vessels and other vessels with variable frequency drive propulsion and variable frequency drive auxiliary services demand further review of the present design process in view of the following:

– Challenge to the system-level electrical stability baseline.

– Challenge to blackout prevention.

– Prevent bolted electric faults-related cascading failure.

– Increase emergency generator size, as well as quantity. (This is a major deviation from present rules and regulations.)

– Install fire-rated cable, and insulated bus pipe for critical applications.

– Understand and manage harmful effects of VFD harmonics.

– MV system grounding managements.

– UPS as a no-break power between blackout and emergency power. (This is a major deviation from present rules and regulations.)

– Train engineering operators in all aspect of shipboard electrical engineering to minimize the control system black-box concept.

– Define performance of vital and redundant auxiliaries at the performance level.

– Clearly define FMEA to better understand the weak links in the electrical system.

– Use a smart reconfiguration control system for vital auxiliaries.

– Monitor system-level electrical ground capacitance.

– Monitor and maintain a baseline electrical system configuration so that any deviation is quickly identified and attended to in a timely manner by a knowledgeable operator.

11.1 VIRTUAL PROTOTYPING FOR ELECTRICAL SYSTEM DESIGN

There are many proven types of software available, such as EDSA, ETAP, SKM, Easy power, Virtual Test Bed (VTB), Simulink, etc. for developing short-circuit analysis, coordination study, power flow calculation, and arc flash analysis. Similar software can be used to develop physics-based prototyping for shipboard power system design. Due to the fact that almost all new designs are customized for specific applications. The traditional proven design may not be suitable for the new requirements, as some

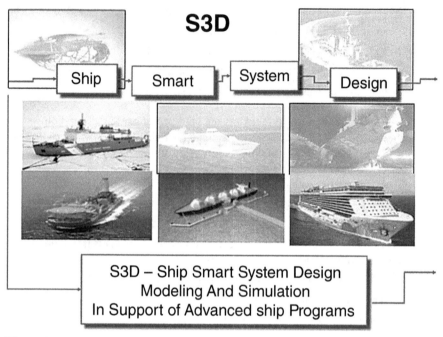

Figure 11.1 S3D Concept Paradigm-1.

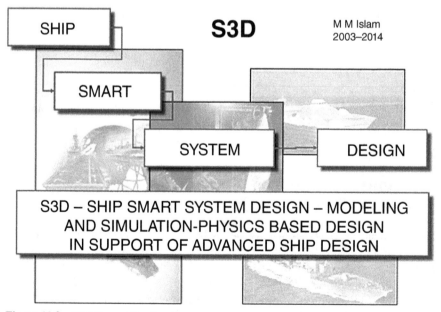

Figure 11.2 S3D Concept Paradigm-2.

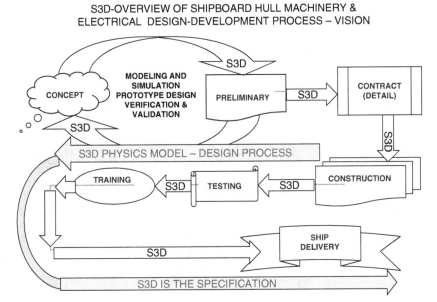

Figure 11.3 Ship Smart System Design (S3D) Paradigm-3.

requirements at the concept-level may not be well defined. Therefore, the same software tools can be used at the concept design phase to develop multiple options under the physics-based simulation. This process will help to bypass the traditional land-based test facility (LBTF) by simulating prototype hardware in the loop design concept. The S3D concept will definitely alleviate the costly time-consuming exercise of

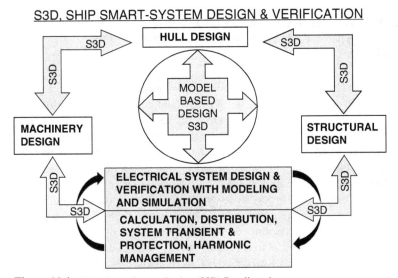

Figure 11.4 Ship Smart System Design (S3D) Paradigm-4.

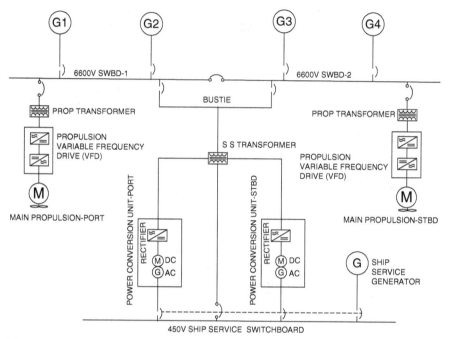

Figure 11.5 Typical Concept EOL for S3D-1.

developing a LBTF, knowing that the LBTF gets obsolete by the time equipment is installed and ready for testing.

11.2 ELECTRICAL POWER SYSTEM SMART SHIP SYSTEM DESIGN FAILURE MODE AND EFFECT ANALYSIS

Smart Ship System Design is model-based systematic analysis of an electrical power system design starting from concept design and continuing through preliminary design and detailed design in view of system-level failure mode and effect analysis (FMEA). However, S3D with FMEA requirements will bring the design to a level where there will be many options to consider and then through S3D fundamentals prioritize the most effective option for the intended application. FMEA should be an independent design body to analyze the design and develop point of common coupling where simulation will be necessary to develop compliance issues, requirement issues, and safety issues. The requirement is to demonstrate the single point failure issues at the point of common coupling of critical systems. Some of those single point failure issues must be identified as to the requirements by regulatory bodies. Others will require additional resolution by calculation, by redundancies, and by protection systems.

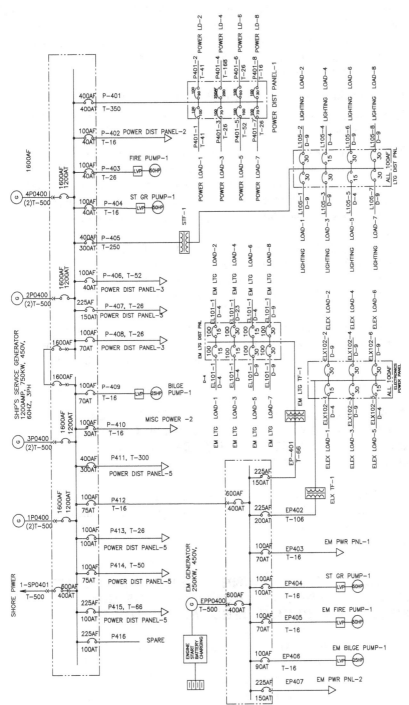

Figure 11.6 Typical Concept EOL For S3D-2.

This EPS S3D is for the early stage of design so that the base of using concept-level design to develop a preliminary design is properly formulated for better understanding of the system requirements from the design to development to procurement, installation, testing, commissioning, training, maintenance, and repair.

To identify the following:

1. Better definition of the requirements by the contract
2. If there are conflicts between the contract requirements and regulatory body requirements
3. Formulate a potential design-related point of common couplings, and single point failures at the critical point of common couplings and alleviate these at the design phase so that the risk of failure is minimized. As the single point failure is addressed at the design level by design modification, by establishing redundancy then the FMEA at the operational level will be easier to deal with, such as:

- Identify the equipment or subsystem, mode of operation;
- Identify potential failure modes and their causes;
- Evaluate the effects on the system of each failure mode;
- Identify measures for eliminating or reducing the risks associated with each failure mode;
- Identify the trials and testing necessary to prove the conclusions; and
- Provide information to the operators and maintainers so that they understand the capabilities and limitations of the system to achieve the best performance.

11.3 MARINE TECHNOLOGY SOCIETY (MTS) GUIDELINES FOR DP VESSEL DESIGN PHILOSOPHY: GUIDELINES FOR MODU DP SYSTEM AND COMMERCIAL SHIPS

The MTS has developed guidelines for DP vessels that are recognized by different authorities having jurisdiction. Therefore the same guidelines are also applicable for all ship design and development. Table 11.1 highlights the MTS requirements and directly relates to commercial ships.

11.4 ADDITIONAL MARINE TECHNOLOGY SOCIETY (MTS) REQUIREMENTS APPLICABLE FOR SHIP DESIGN: USCG RECOGNIZED MTS REQUIREMENTS

(a) MTS3.19 CRITICAL AND NONCRITICAL REDUNDANCY

MTS3.19.1 Class rules require DP systems to be redundant with the primary objective of achieving no loss of position. However, redundancy in itself does not guarantee a particular level of reliability. Loss of fault tolerance could cause operational issues impacting the industrial mission of the

Table 11.1 Redundant System Considerations MODU and Commercial Ships

MTS for MODU DP System	Commercial Ship Equivalant
MTS3.18 MULTIPLE POWER PLANT CONFIGURATIONS MTS3.18.1 Diesel electric plant design should incorporate configuration flexibility to cope with equipment unavailability (e.g., failures or equipment taken down for maintenance). However, it is important that the effect of such reconfigurations are understood as some may not be redundant. Major configurations should be identified and analyzed in the vessel's DP system FMEA to prove the DP system remains redundant. Fault tolerance of configurations should be made visible and understood by the crew. Where there is configuration flexibility in the design, the Critical Activity Mode of Operation (CAMO) should be clearly defined in addition to other Task Appropriate Modes (TAM) for use on DP with any additional risks made visible. For example, some task appropriate modes may rely more heavily on protective functions than others.	Same as MTS. However, Critical Activity Mode of Operation (CAMO) should be clearly defined as the shipboard electrical propulsion plant related CAMO will be different. For shipboard power system, environmental condition is different than the MODU. Adapt S3D concept for shipboard FMEA-based redundancy.
MTS3.19 CRITICAL AND NONCRITICAL REDUNDANCY .MTS3.19.1 Class rules require DP systems to be redundant with the primary objective of achieving no loss of position. However, redundancy in itself does not guarantee a particular level of reliability. Loss of fault tolerance could cause operational issues impacting the industrial mission of the vessel. Where aspects of the design are identified as being of lower reliability or there is a need to ensure higher availability it may be beneficial to provide redundancy over and above that required to meet class requirements. Critical redundancy is defined as equipment required to ensure the vessel is single fault tolerant. To remove such equipment would either remove the DP system's fault tolerance entirely or reduce its post failure DP capability.	For shipboard system, classifications are vital, non-vital, essential, nonessential as defined by rules and regulations. The MTS critical and noncritical redundancies are to be carefully redefined for specific applications.
MTS 9.2.2: Harmonics can also be related to commutation notches from large rectifiers and contribute to overheating of service transformers. High levels of harmonic distortion can have undesirable effects including failure of generator synchronizers, failure of control systems, noisy operation, overheating of machines and failure of ballasts in fluorescent lighting. Harmonics are often a problem in large diesel electric power plants. Measures such as phase shifting transformers, active front end rectifiers, and phase shifting transformers are used to reduce harmonic distortion to acceptable levels. Harmonic filters can be unreliable. There have been known failures on DP vessels leading to severe short circuit faults and associated voltage dips and consequences. Harmonic studies should be carried out to determine the worst case levels of harmonics in the intact condition and following the worst case failure of any harmonic reduction measures.	Same for ships.

(cont.)

Table 11.1   (*Cont.*)

MTS for MODU DP System	Commercial Ship Equivalant
The system should be designed such that post failure levels of harmonics remain within acceptable levels or the power plant should be designed to operate at higher harmonic levels without malfunction. Levels of harmonic distortion should be continuously monitored by the vessel management system and unacceptable levels should initiate an alarm.	
Vessels that operate with the main switchboard bus-ties open should only experience harmonics related to failure on one power system. However, some power system use phase shifting transformers with a different vector group on each side of the main bus-ties to achieve additional harmonic cancellation. It is important that the harmonics remain within acceptable levels when the bus-ties are opened and this harmonic cancellation effect is removed.	
Some types of diesel electric power systems use phase shifting transformer between two bus sections to create a phase shift between the two power system voltage waveforms. When one power systems fails the 12-pulse rectifiers revert to 6-pulse operation. It is important to confirm that operation can continue for as long as required with the higher levels of distortion. Systems using this method should be designed for continuous operation in all defined bus configurations.	
MTS9.2.8: Transient stability: Parallel generators are held in synchronism by the synchronizing torque developed from the bus voltage at the generator's terminals. During a severe short circuit fault the terminal voltage may drop close to zero causing generators to lose synchronism with each other.	
Similar conditions may occur because of the crash synchronization of a generator, or two power systems. Inadvertent connection of a stopped generator may also cause severe disruption. In marine power systems the generators are usually so closely coupled that the plant re-stabilizes when the short circuit has been cleared by the protection. A study should be performed to confirm this.	
MTS9.7.2: Arc detection: Arc detection by optical means or by pressure wave detection has becomes a popular method of bus bar protection for high voltage marine power systems. Arc detection offers advantage of very fast isolation of the fault. It does not depend on detecting the fault current. It does not require coordination with other protection as it positively identifies the location of the fault. It may be supplemented by over current protection to cover the possibility of a short circuit occurring without an accompanying arc.	

vessel. Where aspects of the design are identified as being of lower reliability or there is a need to ensure higher availability it may be beneficial to provide redundancy over and above that required to meet class requirements.

Critical redundancy is defined as equipment required to ensure the vessel is single fault tolerant. To remove such equipment would either remove the DP system's fault tolerance entirely or reduce its post failure DP capability. Noncritical redundancy is equipment intended to provide greater availability.

MTS3.19.2 If redundant elements are highly reliable, there is no need for non critical redundancy but it can be usefully applied to allow maintenance or in cases where it is uneconomical or impractical to increase the reliability further.

(b) MTS3.20 AUTONOMY AND DECENTRALIZATION

MTS3.20.1 Modern DP vessels are complex machines with several layers of automation. Experience suggests that there are benefits to be derived from making generators and thrusters independent in the provision of auxiliary support services and control functions. Designs should be resistant to internal and external common cause and common mode failures. Designs in which the control function has been decentralized are considered to be more fault tolerant. In such designs, each major item of machinery is responsible for making itself ready for operation and ensuring that all necessary services are online. In general, control system failure effects are less likely to exceed loss of the associated engine or thruster. It can be more difficult to prove that the effects of failures in centralized systems do not exceed the worst case failure design intent. This is an important consideration when choosing a control system topology for fault tolerant systems. There is still a requirement for a remote control system in decentralized designs but the functions of this control layer are limited to scheduling and remote manual control.

(c) MTS PROTECTION SYSTEM

    (i) MTS9.7.3: Over current detection: This is the most basic form of protection and is applied at all levels in the power distribution systems for short circuit and over load protection. Over current can be detected by current transformers, fuses, magnetic over current or bi-metal strips with heating coils. At the main power distribution levels 'protection-class' current transformers are used to provide digital relays with a signal representing the line current. Various current versus time curves are used to produce the required degree of coordination with other over current protection upstream and down stream. Note: Protection class CTs may not provide the degree of accuracy required for instrument applications.

    (ii) MTS9.7.4: Differential protection: Differential protection is a form of over current protection based on summing the currents entering and leaving a node such as a switchboard, bus bar or a generator winding. Current transformers are used to monitor the current entering and leaving the zone to be protected. Provided there is no fault path within the zone the currents will sum to zero. If a fault occurs this will no longer be true and a difference signal will be generated operating the over current trip on the circuit breaker. Differential protection can be used to create zones around individual bus sections in a multi-split redundancy concept connected as a ring. With this arrangement only the faulty bus section is tripped and all other bus sections remain connected. This has advantages if some of the bus sections do not have a generator connected. Differential protection schemes can have problems with high levels of through-fault current. That is current passing through a healthy zone on its way to a fault in some other zone. There have been problems with healthy zones tripping causing failure effects exceeding WCFDI. It is for this reason that some designers favor arc protection for this application.

The effectiveness of differential protection for bus bar applications is difficult to establish conclusively without conducting short circuit testing. Differential protection is almost universally applied for the protection of generator windings on machines above about 1.5MVA.

(iii) MTS9.7.5: Directional over current protection: Directional over current protection is sometimes applied for bus-bar protection. It is less expensive than differential, due to the reduced number of current transformers required to define a protection zone. Directional over current generally cannot be used with ring configurations as it depends on blocking the upstream circuit breaker from tripping.

(iv) MTS9.7.6: Earth (Ground) fault protection: The size of the power distribution system and the maximum prospective earth fault current influences the type of earth fault protection specified for marine system. Low voltage marine power systems are often designed as un-intentionally earthed systems where the power system has no direct connection or reference to earth (vessel's hull). On these systems, earth faults are typically indicated by earth fault lamps or meters connected from each line to earth. Intentional earth impedance should be considered in the case of high voltage systems. High resistance earthing (grounding) of various types is generally employed. All power systems are referenced to earth by way of the distributed capacitance of cables and windings. A significant earth fault current can flow even in unintentionally earthed HV systems. The intentional earth impedance adds to the system charging current when an earth fault occurs and should be sized to provide an earth fault current three times that which would flow as a result of the capacitive charging current. This provides well defined current paths for protection purposes. Earth fault protection for the main power system is sometimes based solely on time grading. The relay in the earthing (grounding) resistor or earthing (grounding) transformers for each bus will detect an earth fault at any point in the plant not isolated by a transformer. Earth fault protection in the feeders is used to isolate a fault in a consumer. If the earth fault persists after the tripping time of the feeder the fault is assumed to be in the generators or on the bus-bars itself. At this point the protection driven from the neutral earthing (grounding) transformers will trip the main bus-ties to limit the earth fault to one bus or the other. Whichever neutral earthing (grounding) transformer continues to detect an earth fault will then trip all generators connected to that bus. Losing a whole bus due to an earth fault in one generator is unnecessarily severe. Design should consider adding restricted earth fault protection to the generators.

(v) MTS9.7.7: Over under voltage: This protection element is often a class requirement. It assists in preventing equipment damage but does not contribute to redundancy concept directly. There should be other protective functions to prevent the power plant reaching the point at which this protection operates. Over / under voltage protection is not selective and blackout is the likely outcome. To prevent blackout in common power systems (closed bus), design should provide other protective functions which detect the onset of the voltage excursion and divide the common power system into independent power systems or isolate the sources of the fault before healthy generators are tripped (for example a faulty generator). Operating the power system as two or more independent power system (bus-ties open) provides protection against this fault.

(vi) MTS9.7.8: Over under frequency: Under frequency can be caused by system overload and there must be means of preventing the power plant reaching this condition. Such functions are normally found in the DP control system, power management system, thruster drives and other large drives. Over frequency can be caused by a governor failing to the full fuel condition. This will cause a severe load sharing imbalance which can drive up the bus frequency to the point where several healthy generators trip on over frequency or reverse power. The failure scenarios are similar to those for over and under voltage as described above.

(vii) MTS9.7.9: Reverse power: This protective function is applied to prevent a diesel generator that has lost power from becoming an unacceptable burden on other generators operating in parallel. If a generator with a fuel supply problem sheds the load it is carrying it will be motored by other generators. The power required to motor the faulty generator adds to the load on the healthy generators. Although the reverse power trip is a useful function, it makes healthy generators vulnerable to being forced to trip on reverse power if a faulty generator takes all the load. In this failure scenario the healthy generators all trip on reverse power and the faulty set trips on some other protective function leading to blackout. Vessels operating their power plant as a common power system should have a means to detect the onset of a generator fault which could have this effect and either subdivide the power plant into independent power systems or trip the generator that is creating the problem. Operating the power plant as two or more independent power systems (bus-ties open) provides protection against this type of failure.

(For additional details refer to MTS guide)

11.5 CONDITION-BASED MAINTENANCE

The complex nature of shipboard electrical system failures may create unacceptable situations. Some failures can be categorized as catastrophic due to the fact that the protection functions may not be adequate. The S3D concept can very well support proven technologies, such as condition-based maintenance (CBM). Regardless of the complexity of the electrical and electronic systems, condition-based health monitoring will guide the design engineers as well as the operators to better understanding of the system, so that the system's reliability is improved.

11.6 FMEA OBJECTIVES: S3D CONCEPT

FMEA is a systematic analysis of the systems to whatever level of detail is required to demonstrate that no single failure will cause an undesired event.

This is to identify potential design and process failures before they occur and to minimize the risk of failure by either proposing design changes or, if these cannot be formulated, proposing operational procedures. Essentially the FMEA is to:

- Identify potential failure modes and their causes;
- Evaluate the effects on the system of each failure mode;
- Identify measures for eliminating or reducing the risks associated with each failure mode;
- Identify trials and testing necessary to prove the conclusions; and
- Provide information to the operators so that they understand the system capabilities and limitations to achieve best performance.

The S3D design tool can be an excellent undertaking to minimize single point failure of critical systems.

Remarks: The S3D concept was initiated by the author during shipbuilding industry day adapting "All Electric Power System" (AEPS) concept.

11.7 ADDITIONAL S3D PROCESS SAFETY FEATURES

- DSM   Design Safety Management
- ORM   Operational Risk Management

- LOPA   Layer of Protection Analysis
- SIS   Safety Instrumented System
- SIF   Safety Instrumented Function
- SIL   Safety Integrity Level
- RRF   Risk Reduction Factor
- PHA   Process Hazard Analysis
- HAZOP Hazards and Operability
- FMEA Failure Mode and Effect Analysis
- QFA   Qualitative Failure Analysis
- DVTP Design Verification Test Procedure
- BLACK-OUT Prevention
- RIDE THROUGH FEATURE

Chapter 12

Electrical Safety and Arc Flash Analysis

12.0 INTRODUCTION

Electricity is very dangerous, regardless of the use. Nowadays electrical cords and jumpers are everywhere with many different combinations and options to use single-phase, three-phase, two-wire, three-wire, and four-wire systems. At the electrical termination point, users are concerned mostly for the availability of power. However, there are complexities beyond those plug and connection points. It is very important to educate ourselves about electrical safety basics, and then take precautions so that the deadly behavior of electricity is better understood, to avoid serious consequences.

All electrical systems have the potential to cause harm. Electrical conductors are materials that allow the movement of electricity through them. Most metals are conductors. A ship's hull is a conductor. The human body is also a conductor.

Electric current cannot exist without an unbroken path to and from the conductor. Electricity will form a "path" or "loop." When you plug in a device (e.g., a power tool), the electricity takes the easiest path from the plug-in, to the tool, and back to the power source. This is also known as creating or completing an electrical circuit.

12.1 INJURIES RESULT FROM ELECTRICAL-CURRENT SHORTS

People are injured when they become part of the electrical circuit. Humans are more conductive than the earth (the ground we stand on), which means if there is no other easy path, electricity will try to flow through our bodies.

Shipboard Power Systems Design and Verification Fundamentals, First Edition. Mohammed M. Islam.
© 2018 the Institute of Electrical and Electronics Engineers, Inc. Published 2018 by John Wiley & Sons, Inc.

There are different types of injuries: electrocution (fatal), electric shock, burns, and falls. These injuries can happen in various ways:

a. Direct contact with exposed energized conductors or circuit parts. When electrical current travels through our bodies, it can interfere with the normal electrical signals between the brain and our muscles (e.g., the heart may stop beating properly, breathing may stop, or muscles may spasm).

b. When the electricity arcs (or jumps) from an exposed energized conductor or circuit part (e.g., overhead power lines) through a gas (such as air) to a person who is grounded (that would provide an alternative route to the ground for the electrical current).

c. Thermal burns including burns from heat generated by an electric arc, and flame burns from materials that catch on fire from heating or ignition by electrical currents or an electric arc flash. Contact burns from being shocked can burn internal tissues while leaving only very small injuries on the outside of the skin.

d. Thermal burns from the heat radiated from an electric arc flash. Ultraviolet (UV) and infrared (IR) light emitted from the arc flash can also cause damage to the eyes.

An arc blast can release a potential pressure wave. This wave can cause physical injuries, collapse the lungs, or create noise that can damage hearing.

Muscle contractions, or a startle reaction, can cause a person to fall from a ladder, scaffold or aerial bucket. The fall can cause serious injuries.

12.2 GENERAL SAFETY TIPS FOR WORKING WITH OR NEAR ELECTRICITY

a. Inspect portable cord-and-plug connected equipment, extension cords, power bars, and electrical fittings for damage or wear before each use. Repair or replace damaged equipment immediately.

b. Always tape extension cords to walls or floors when necessary. Nails and staples can damage extension cords, causing fire and shock hazards.

c. Use extension cords or equipment rated for the level of amperage or wattage that you are using.

d. Always use the correct size fuse. Replacing a fuse with one of a larger size can cause excessive currents in the wiring and possibly start a fire.

e. Be aware that unusually warm or hot outlets may be a sign that unsafe wiring conditions exist. Unplug any cords or extension cords to these outlets and do not use until a qualified electrician has checked the wiring.

f. Always use ladders made with nonconductive side rails (e.g., fiberglass) when working with or near electricity or power lines.

12.3 ARC FLASH BASICS

Arc flash analysis is not mandatory for shipboard power systems. The fundamentals related to shipboard arc flash analysis with fault current analyses are presented with software-driven modeling for arc flash analysis.

Traditional shipboard power systems have been changed to higher voltage with high-power generation and complex distribution systems such as electrical propulsion drive with adjustable speed drive applications. The arc flash incidents onboard ships demand arc flash analysis along with power system protection device coordination study. Although National Fire Protection Agency (NFPA) 70E specifically deals with industrial power system arc flash, there are requirements that are equally applicable to shipboard power systems. In view of National Electric Code (NEC) requirements, and IEEE 1584 recommendations, the performance of shipboard power system arc flash analysis is strongly recommended for all new shipboard power system designs as well as for existing ships.

Shipboard electrical power system requirements have been changed due to the proliferation of electric propulsion, and electric drive auxiliaries. Shipboard power generation and distribution requirements are being challenged due to limited space in the ship as well as the dynamic behavior of ships in motion, saltwater contact with the hull, as well as atmospheric moisture. Many shipboard electrical accidents are attributed to electrical arc flash situations starting with phase to ground and then phase to phase leading to bolted fault. Shipboard electrical system arc flash analysis has become urgently needed for the safety of the equipment as well as the safety of the operators.

The efficient way to mitigate arc flash incident energy onboard ship is to detect the fault as soon as it happens and clear the fault as fast as possible. There are ways to detect and clear the arcing fault such as:

(1) Use fiber optic light sensors.

(2) Modify the existing settings of protective devices.

(3) Apply new technologies that have been developed to detect and clear the arcing fault instantly during an arc flash incident.

(4) Take advantage of alternative protection schemes such as differential protection and zone interlocking, since these types of schemes clear the fault instantaneously.

For industrial power systems, the arc flash requirements are in OSHA 29 Code of Federal Regulations (CFR) Part 1910 Subpart S, NFPA-70-2012, National Electrical Code, NFPA-70E-2012, Standard for Electrical Safety Requirements for Employees workplaces, and IEEE Std 1584-2002 Guide for Performing Arc Flash Hazard Calculations.

Shipboard power system fault current analysis is done using many different methods such as the USCG guide, IEEE guide, NASEA design data, IEC, and a combination of smart software. Some of those are presented to provide users with many choices of applications. However, it is recommended that arc flash study be performed for each ship.

12.4 FUNDAMENTALS OF ELECTRICAL ARC AND ARC FLASH

Shipboard arc flash is one of the most dangerous workplace hazards. The space area for electrical equipment is very limited onboard ship. A shipboard electrical system is ungrounded. An arc flash incident occurs when electric current passes between two conducting metals through ionized air. When this arc flash phenomenon happens, heat (incident energy) is released, which can burn human skin and set clothing on fire. The arc energy depends on the arcing current and duration. Besides the arcing heat, electrical arc flash produces a high-pressure blast that can produce molten metal and shrapnel, as well as sounds.

Equations developed by electrical safety standards (IEEE 1584 and NFPA 70E) to calculate the potential incident energy caused by an arcing fault indicates that the faster the arcing fault is cleared, the lower the incident energy. Therefore, one of the best and most efficient ways to mitigate arc flash incident energy is to clear the arcing fault as fast as possible.

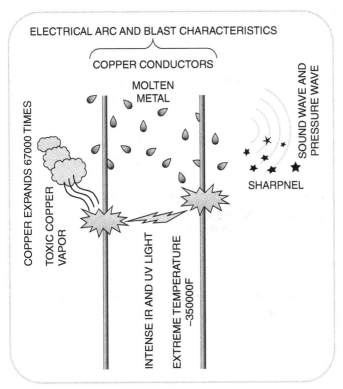

Figure 12.1 Electric Arcing Basics.

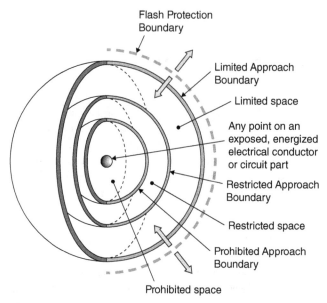

Figure 12.2 Arcing Approach Limit.

The shipboard power system design and development process must include arc prevention, protection, and mitigation systems. There are various ways to identify the arcing fault, minimize, and then clear it. One way is by simply modifying the existing settings of protective devices. Another way is to apply new technologies such as light sensors to detect arc and then clear the arcing fault instantly during an arc flash incident. Finally, another way is by taking advantage of alternative protection schemes such as differential protection and zone interlocking, since these types of schemes clear the fault instantaneously.

The easiest way to reduce the clearing time of arc fault is by reducing the instantaneous setting or the short time pickup (STPU) setting of a protective device.

12.5 DEFINITIONS RELATED TO ARC FLASH (DERIVED FROM NFPA 70E NEC, NFPA 70E, AND IEEE STD 1580 FOR SHIPBOARD ELECTRICAL INSTALLATIONS)

(a) **Arc Flash boundary:** *(NFPA-70E) When an arc flash hazard exists, an approach limit at a distance from a prospective arc source within which a person could receive a second degree burn if an electrical arc flash were to occur.*

(b) **Limited Approach Boundary:** *(NFPA-70E) An approach limit at a distance from an exposed energized electrical conductor to circuit part within which a shock hazard exists. (It is the closest distance an unqualified person can approach, unless accompanied by a qualified person.)*

(c) **Restricted Approach Boundary:** *(NFPA-70E) An approach limit at a distance from an exposed energized electrical conductor to circuit part within which is an increased risk of a shock due to electrical arc over combined with inadvertent movement, for personnel working in close proximity to the closest distance to exposed energized electrical conductor or circuit part a qualified person can approach without proper PPE and tool.*

(d) **Prohibited Approach Boundary:** *(NFPA-70E) An approach limit at a distance from an exposed energized electrical conductor to circuit part within which work is considered the same as making contact with electrical conductor or circuit part. (It is the minimum approach distance to exposed energized electrical conductor or circuit part to prevent flashover or arcing.)*

(e) **Incident Energy:** The incident energy is a measure of thermal energy at a working distance from an arc fault. The unit of incident energy is cal/cm^2 or joule/cm^2.

(f) **PPE:** Personal Protective Equipment

12.6 CAUSES OF ELECTRIC ARC

Electric arc can be initiated by many factors, such as:

(a) Dust and other impurities:
  (1) Dust and other impurities on the insulating surface can provide a path for current, allowing it to flashover and create arc discharge across the surface.
  (2) Fumes or vapors of chemicals can reduce the breakdown voltage of air and cause arc flash.

(b) Shipboard Corrosion: Corrosion is a major factor onboard ship due to a salt-laden corrosive atmosphere. Corrosion: Corrosion of electrical equipment parts can provide impurities on the insulating surface and cable connections. Corrosion weakens and reduces the contact between conductor terminals, with an increase in contact resistance through oxidation. Usually heat is generated on the contacts creating sparks.

(c) Shipboard condensation:
  Shipboard condensation is a major factor for insulation failure and contact oxidation.

(d) Inadvertent contact to the live circuit.

(e) Loose electrical connection.

12.7 INCIDENT ENERGY

This is the amount of thermal incident energy to which personnel could be exposed at working distance during an electrical arc event. Incident energy is measured in joules per centimeter squared (J/cm^2) or calories per centimeter squared (cal/cm^2).

The unit of incident energy is cal/cm^2 or joule/cm^2

One cal/cm^2 = 4.184 joule/cm^2

One joule/cm^2 = 0.239 cal/cm^2

12.8 INCIDENT ENERGY AT ARC FLASH PROTECTION BOUNDARY

Enter a value in cal/cm^2 to determine the arc flash protection boundary (FPB) distance at that Incident Energy.

The Incident Energy of 1.2 cal/cm^2 for bare skin is used in solving equation for Flash Protection Boundary in IEEE 1584 Guide for Performing Arc Flash Hazard Calculations. However, the Guide equation for Flash Protection Boundary can be solved with other incident energy levels as well such as the rating of proposed personal protective equipment (PPE). A minimum of 0.3 cal/cm^2, and a maximum of 3 cal/cm^2 are also used.

12.9 THE FLASH PROTECTION BOUNDARY

The flash protection boundary is an approach limit at a distance from exposed live parts or enclosed live parts if operation, manipulation, or testing of equipment creates a potential flash hazard, within which a person could receive a second degree burn if an electrical **arc flash** were to occur. A worker entering the flash protection boundary must be qualified and must be wearing appropriate PPE. The flash protection boundary is required to be calculated by NFPA 70E.

12.10 ELECTRICAL HAZARDS: ARC FLASH WITH ASSOCIATED BLAST AND SHOCK

Abnormal conditions on energized electrical equipment will increase the potential risk of exposure to workers to the electrical hazards of arc flash with associated blast and shock. The risk exposure will increase depending on the level of interaction with the energized electrical equipment. Some abnormal conditions that can increase the risk of exposure:

– Doors open on energized equipment, lower risk when hinged doors.

– Doors open on energized equipment, higher risk when bolted covers have to be removed.

– Overcurrent conditions where the gap between conductors is compromised.

– Energized electrical equipment integrity negatively impacted by the operating environment.

The level of risk will also be influenced by:

– Level of human interaction related to the work task and any changes of the state of the electrical equipment when the hinged door is open or covers are removed and the electrical equipment is energized.

– Lack of NEC compliance.

– Lack of maintenance.

– Effects of aging.

In order to minimize the risk of exposure, appropriate preventive and protective control measures should be reviewed and implemented with the first priority to eliminate the hazard if at all possible by de-energizing. If the hazard cannot be eliminated or limited then the risk of exposure needs to be decreased to acceptable levels.

With respect to the electrical hazards is it important to understand how they are defined. As per CSA Z462 and NFPA 70E the following definitions are provided:

12.11 SHOCK HAZARD

A dangerous condition associated with the possible release of energy caused by contact with or approach to energized electrical conductors or circuit parts.

A prioritized list for preventive and protective control measures should be used for the mitigation or reduction of exposure to workplace hazards. These same principles can and should be applied to the electrical hazards of shock and arc flash. Action taken should be consistent with the following prioritized preventive and protective control measures:

– Eliminate the hazard, de-energize is the first choice.

– Reduce the risk by engineering design (e.g., engineering solutions, equipment solutions, "Safety by Design"). Ensure that adequate electrical equipment maintenance is performed and at an acceptable frequency.

– Use safer work systems that increase awareness of potential hazards (e.g., apply safeguards such as signage, barriers, etc.).

– Implement administrative controls (e.g., training and procedures); and

– Use electrical epecific PPE, tools and equipment as a last line of defense and ensure these are appropriately used and maintained.

NFPA 70E identifies Shock Approach Boundaries that are to be applied when a shock hazard risk exists. They are the Limited Approach Boundary, Restricted Approach Boundary, and Prohibited Approach Boundary. The shock risk increases based on the distance being reduced from the Limited to Prohibited Approach boundaries. Only qualified persons are allowed inside these boundaries, and specifically the Restricted and Prohibited Approach boundaries. See Figure 1 Boundaries for Arc Flash Protection and Shock: Approach Limits for a graphical representation of the Shock Approach Boundaries.

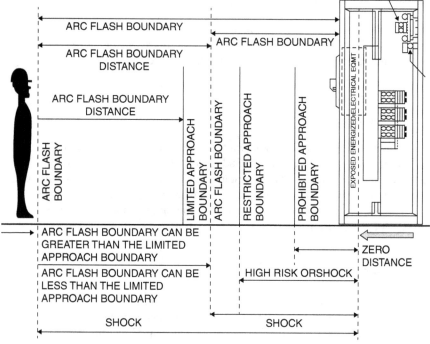

Figure 12.3   Arc Flash Boundary Sample-1.

12.12 HAZARD/RISK CATEGORIES (DERIVED FROM NFPE-70E)

12.12.1 Hazard/Risk Category: Description (HRC-0)

Table 12.1   Hazard/Risk Categories (HRC-0)

HAZARD/RISK CATEGORY HRC-0	Cotton undergarment	Arc rated
	Long sleeve shirt (natural fiber)	Arc rated
	Long pant (natural fiber)	Arc rated
	Safety sun glass	
	Hearing protection	
	Insulated glove	

12.12.2 Hazard/Risk Category: Description (HRC-1)

Table 12.2   Hazard/Risk Categories (HRC-1)

HAZARD/RISK CATEGORY HRC-1	Cotton undergarment	Arc rated
	Long sleeve shirt (natural fiber)	Arc rated
	Long pant (natural fiber)	Arc rated
	Safety sun glass	
	Hearing protection	
	Insulated glove	

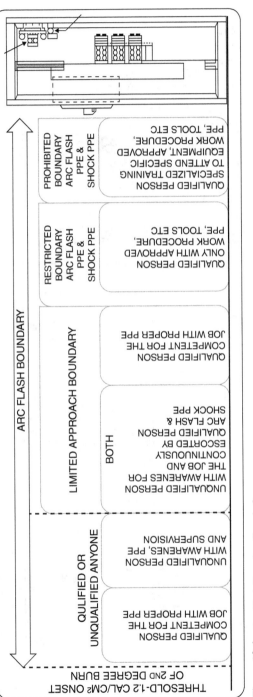

Figure 12.4 Arc Flash Boundary Sample-2.

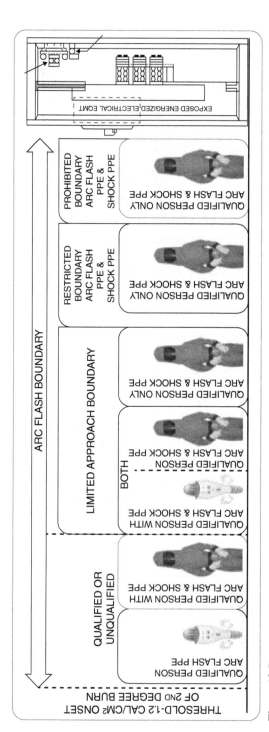

Figure 12.5 Arc Flash Boundary Sample-3.

12.12.3 Hazard/Risk Category: Description (HRC-2)

Table 12.3   Hazard/Risk Categories (HRC-2)

HAZARD/RISK CATEGORY HRC-2	Cotton undergarment	Arc rated
	Long sleeve shirt (natural fiber)	Arc rated
	Long pant (natural fiber)	Arc rated
	Safety sun glass	
	Hearing protection	
	Insulated glove	

12.12.4 Hazard/Risk Category: Description (HRC-3)

Table 12.4   Hazard/Risk Categories (HRC-3)

HAZARD/RISK CATEGORY HRC-3	Cotton undergarment	Arc rated
	Long sleeve shirt (natural fiber)	Arc rated
	Long pant (natural fiber)	Arc rated
	Safety sun glass	
	Hearing protection	
	Insulated glove	

12.12.5 Hazard/Risk Category: Description (HRC-4)

Table 12.5   Hazard/Risk Categories (HRC-4)

HAZARD/RISK CATEGORY HRC-4	Cotton undergarment	Arc rated
	Long sleeve shirt (natural fiber)	Arc rated
	Long pant (natural fiber)	Arc rated
	Safety sun glass	
	Hearing protection	
	Insulated glove	

12.13 SHIPBOARD ELECTRICAL SAFETY COMPLIANCE CHART PER NFPA 70E 2012 TABLE 130.7.C.9

Table 12.6   Task Performed on Shipboard Energized Equipent: LV Switchboard and Panelboard with Power Circuit Breaker, Molded Case and Insulated Case Circuit Breakers

Item	Task	HRC Level	Remarks
1	Noncontact inspection outside restricted approach boundary	1	
2	Circuit beaker operation with cover on	0	
3	Circuit beaker operation with cover off	1	
4	Work on energized electrical conductor and circuit parts	2	

TASK-RELATED HAZARD/RISK CATEGORY (HRC) CHART

– Unqualified workers can be inside the Limited Approach Boundary, but must be supervised by a Qualified Person.

– Work within the Restricted Approach Boundary will have a higher risk of shock exposure due to inadvertent or direct contact with the exposed energized electrical conductors or circuit parts. A primary protection strategy is to "guard" or "insulate" exposed conductors or circuit parts which eliminates the risk of shock. If this cannot be done recommend the use of rubber insulating gloves with leather protectors as the primary shock PPE that can used by a worker. The use of insulated and insulating hand tools would be a secondary control measure. Of course, de-energizing would be the first choice before any work is performed on the conductors or circuit parts.

The arc flash hazard study shall consider all operating scenarios during normal conditions alternate operations, emergency power conditions, and any other operations, which could result in maximum arc flash hazard. The label shall list the maximum incidental energy calculated and the Scenario number and description on the label.

12.14 ARC FLASH: OSHA REQUIREMENTS (29 CFR 1910.333)

Safety-related work practice shall be employed to prevent electric shock or other injuries resulting from either direct or indirect electric contacts.

"Live electric parts to which an employee may be exposed shall be de-energized before the employee works on or near them, unless the employer can demonstrate that de-energizing introduces additional or increased hazards or is infeasible."

(Notes for ship: If there is a redundant equipment available, de-energize the faulty equipment to work on dead system. If there is no redundant system or equipment then chief engineer to decide on working on live equipment. The flash analysis must be followed to ensure all safety measures.)

(Working on energized equipment is always a risk, therefore, the fundamental requirement is to de-energize equipment prior to work.)

12.15 ARC FLASH: NATIONAL ELECTRICAL CODE (NEC) REQUIREMENTS

NEC-2005 -110.16 Flash Protection. Shipboard electrical Switchboards, panel boards, industrial control panels, and motor control centers require examination, adjustment, servicing, or maintenance while energized shall be field marked to warn qualified persons of potential electric arc flash hazards. The marking shall be located so as to be clearly visible to qualified persons before examination, adjustment, servicing, or maintenance of the equipment

NEC FPN 110.16 No. 1: NFPA 70E-2012, *Standard for Electrical Safety in the Workplace,* provides assistance in determining severity of potential exposure, planning safe work practices, and selecting personal protective equipment.

NEC 110.16 FPN No. 2: ANSI Z535.4-1998, *Product Safety Signs and Labels,* provides guidelines for the design of safety signs and labels for application to products.

NEC 2012-II. 600 V, Nominal, or Less

110.26 Spaces About Electrical Equipment. Sufficient access and working space shall be provided and maintained about all electric equipment to permit ready and safe operation and maintenance of such equipment. Enclosures housing electrical apparatus that are controlled by a lock(s) shall be considered accessible to qualified persons.

(A) Working Space. Working space for equipment operating at 600 V, nominal, or less to ground and likely to require examination, adjustment, servicing, or maintenance while energized shall comply with the dimensions of 110.26(A)(1), (A)(2), and (A)(3) or as required or permitted elsewhere in this *Code.*

(1) Depth of Working Space. The depth of the working space in the direction of live parts shall not be less than that specified in Table 110.26(A)(1)(1) unless the requirements of 110.26(A)(1)(a), (A)(1)(b), or (A)(1)(c) are met. Distances shall be measured from the exposed live parts or from the enclosure or opening if the live parts are enclosed.

Table 110.26(A)(1) Working Spaces

NFPA-70E STD FOR ELECTRICAL SAFETY IN WORK PLACES:

110-8B-1.b: Arc Flash Hazard Analysis-An arc Flash Hazard analysis shall determine the ARC Flash Protection boundary and the personal protection equipment that people within the Arc Flash Protection Boundary shall use.

FPN: See 130.3 for the requirements of conducting an arc flash analysis.

12.16 ARC FLASH: NFPA 70E 2012 REQUIREMENTS

(a) Article 100 FPN: Arc-rated clothing indicates that it has been tested for exposure to an electric arc. Flame-resistant (FR) clothing without an arc rating has not ben tested for exposure to an electric arc.

All arc-rated clothing is required to include a visibly marked Arc Thermal Performance Exposure Value (ATPV) rating.

The expression "Arc rates" is the replacement of "flame-resistant."

(b) Sec. 110.2(D): The employer shall determine, through regular supervision or through inspections conducted on at least an annual basis that each employee is complying with the safety-related work practices as required by NFPA-70E. (2012 version changed from no specific frequency to annual)

(c) Section 110.3: The Electrical Safety Program shall be audited to verify the principles and procedures of the electrical safety program are in compliance with NFPA-70E. The frequency of the audit shall not exceed 3 years. (This was left to the program manager)

(d) Section 130.2: Electrically Safe Working Conditions. Energized electrical conductors and circuit parts to which an employee might be exposed shall be put into an electrically safe work condition before an employee performs work if either of the following conditions exists:

(1) The employee is within the limited approach boundary.

(2) The employee interacts with equipment where conductors or circuit parts are not exposed, but an increased risk of injury from an exposure to an arc flash hazard exists.

(e) Section 130.2(B)(2): Work permit requirements

(a) Limited approach boundary

(b) Restricted approach boundary

(c) Prohibited approach boundary

(d) Necessary shock personal and other protective equipment to safely perform the assigned task

Section 130.2(B)(5): Flash hazard analysis requirements:

(a) Available incident energy or hazard/risk category (See 130.5)

(b) Necessary personal protective equipment to safely perform the assigned task

(c) Arc boundary

(f) Section 130.5(c): Arc flash labeling. Electrical equipment such as switchboards, panel boards, control panels, and motor controllers, motor control centers adjustment, servicing, or maintenance while energized, shall be field marked with a label containing the following information: (as applicable)

(1) At least one of the following

(a) Available incident energy and the corresponding working distance

(b) Minimum arc rating of the clothing

(c) Required label PPE

(d) Highest Hazard/Risk Category (HRC) for the equipment

(2) Nominal system voltage

(3) Arc flash boundary

(Previous editions simply stated that "equipment shall be field marked with a label containing the available incident energy or required level of PPE.")

(g) New Annex-H Table H.3(a): Provides guidance on the application of NFPA 70E sections related to shock hazard PPE and arc flash PPE

(h) New Table H.4(a) and Table H.4(b): Provides maximum three-phase bolted fault current limits at low-voltage systems (208 V through 690 V) and high-voltage systems (5 kV, 12 kV, and 15 kV respectively), along with circuit breaker fault clearing times for the recommended use of 8 cal/cm[1] and 40 cal/cm PPE in what is termed an "arc -in-a-box" situation. The limitations noted in these tables are based on IEEE 1584 calculation methods.

> FPN: For protective devices operating in the steep portion of their time-current curves of the protective devices, a small change in current causes a big change in operating time. Incident energy is linear with time, so arc current variation may have a big effect on incident energy. The solution is to make two or more arc current and energy calculations: one using one of the calculated expected arc currents and one using one of the reduced arc currents and then lower it by some acceptable percentage.

12.17 ARC FLASH BOUNDARY: NFPA 70E

(a) LIMIT APPROACH—BOUNDARY REQUIREMENT CHANGES IN NFPA 70E 2012 EDITION

Table 12.7 Limit Approach—Boundary Requirement Changes in NFPA 70E 2012 Edition

	Changes From	Changes To	Remarks
1	Flash Protection Boundary	Arc Flash Boundary	
2	Flash Hazard Boundary	Arc Flash Boundary	
3	Flash Boundary	Arc Flash Boundary	
4	Fire Rated (FR)	Arc Rated	Must be tested for ASTM arc rating
5			
6			

[1] The employee interacts with equipment where conductors or circuit parts are not exposed, but an increased risk of injury from an exposure to an arc flash hazard exists.

   (e) Work permit requirement: Section 130.2(B)(2)

   (a) Limited approach boundary

   (b) Restricted approach boundary

   (c) Prohibited approach boundary

   (d) Necessary shock personal and other protective equipment to safely perform the assigned task Section 130.2(B)(5): Flash Hazard analysis requirements:

     (a) Available incident energy or hazard/risk category (See 130.5)

     (b) Necessary personal protective equipment to safely perform the assigned task

     (c) Arc boundary

   (f) Arc Flash Labeling (Section 130.5(c)): Electrical equipment such as switchboards, panel boards, control panels, and motor controllers, motor control centers adjustment, servicing, or maintenance while energized, shall be field marked with a label.

Remarks: Electrical safety program shall identify procedure for working "Limited approach boundary" and "Arc Flash Boundary."

12.18 LOW-VOLTAGE (50 V–1000 V) PROTECTION (NFPA 70E 130.3 (A1))

Voltage Levels Between 50 V and 1000 V: In those cases where detailed arc flash hazard analysis calculations are not performed for systems that are between 50 V and 1000 V, the Arc Flash Protection Boundary shall be 4.0 ft, based on the product of clearing time of 2 cycles (0.033 sec) and the available bolted fault current of 50 kA or any combination not exceeding 100 kA cycles (1667 ampere second). When the product of clearing times and bolted fault current exceeds 100 kA cycles, the Arc Flash Protection Boundary shall be calculated.

12.19 MEDIUM-VOLTAGE (1000 V AND ABOVE) (NFPA 70E 130.3 (A2))

Voltage Levels Above 1000 V: At voltage levels 1000 V, the Arc Flash Boundary shall be the distance at which the incident energy equals 5 J/cm^2

---

[2] The employee interacts with equipment where conductors or circuit parts are not exposed, but an increased risk of injury from an exposure to an arc flash hazard exists.

(e) Work permit requirement: Section 130.2(B)(2)

(a) Limited approach boundary

(b) Restricted approach boundary

(c) Prohibited approach boundary

(d) Necessary shock personal and other protective equipment to safely perform the assigned task Section 130.2(B)(5): Flash Hazard analysis requirements:
  (a) Available incident energy or hazard/risk category (See 130.5)
  (b) Necessary personal protective equipment to safely perform the assigned task
  (c) Arc boundary

(f) Arc Flash Labeling (Section 130.5©): Electrical equipment such as switchboards, panel boards, control panels, and motor controllers, motor control centers adjustment, servicing, or maintenance while energized, shall be field marked with a label containing the following information: (as applicable) (1) At least on of the following

(a) Available incident energy and the corresponding working distance

$(1.2\,\text{cal/cm}^3)$. For situations, where fault-clearing time is equal to or less than 0.1 sec, the Arc Flash Protection Boundary shall be the distance at which the incident energy level equals $6.24\,\text{J/cm}^4$ (1,5 cal/cm).

12.20 ARC FLASH: IEEE 1584 REQUIREMENTS AND GUIDELINES

(a) Develop electrical one-line diagram

(b) Collect power system data

(c) Determine the bolted fault current

(d) Determine the arc fault current

(e) Define arcing duration from the device characteristics

---

[3] The employee interacts with equipment where conductors or circuit parts are not exposed, but an increased risk of injury from an exposure to an arc flash hazard exists.
   (e) Work permit requirement: Section 130.2(B)(2)

   (a) Limited approach boundary

   (b) Restricted approach boundary

   (c) Prohibited approach boundary

   (d) Necessary shock personal and other protective equipment to safely perform the assigned task
      Section 130.2(B)(5): Flash Hazard analysis requirements:
      (a) Available incident energy or hazard/risk category (See 130.5)
      (b) Necessary personal protective equipment to safely perform the assigned task
      (c) Arc boundary

   (f) Arc Flash Labeling (Section 130.5©): Electrical equipment such as switchboards, panel boards, control panels, and motor controllers, motor control centers adjustment, servicing, or maintenance while energized, shall be field marked with a label containing the following information: (as applicable) (1) At least on of the following

  (a) Available incident energy and the corresponding working distance

[4] The employee interacts with equipment where conductors or circuit parts are not exposed, but an increased risk of injury from an exposure to an arc flash hazard exists.
   (e) Work permit requirement: Section 130.2(B)(2)

   (a) Limited approach boundary

   (b) Restricted approach boundary

   (c) Prohibited approach boundary

   (d) Necessary shock personal and other protective equipment to safely perform the assigned task
      Section 130.2(B)(5): Flash Hazard analysis requirements:
      (a) Available incident energy or hazard/risk category (See 130.5)
      (b) Necessary personal protective equipment to safely perform the assigned task
      (c) Arc boundary

   (f) Arc Flash Labeling (Section 130.5©): Electrical equipment such as switchboards, panel boards, control panels, and motor controllers, motor control centers adjustment, servicing, or maintenance while energized, shall be field marked with a label.

Table 12.8   PPE Hazard/Risk Classification as per NFPA 70E-2000 (Medium Voltage)

PPE Catagory	Energy Level	Typical PPE Examples	Remarks
0	N/A		Ppe must be approved for arc rating
1	5 cal/cm^2		
2	8 cal/cm^2		
3	25 cal/cm^2		
4	40 cal/cm^2		

(f) Record system voltage and equipment class

(g) Determine working distance

(h) Determine incident energy for each work location in the study

(i) Determine flash protection boundary for each work location in the study

12.21   ARC FLASH: CIRCUIT BREAKER TIME CURRECT COORDINATION – OVERVIEW

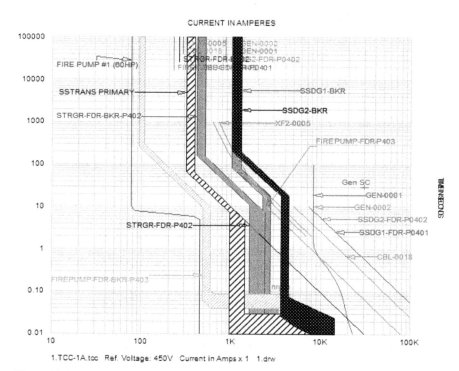

Figure 12.6   Protective Device Coordination for Arc Flash.

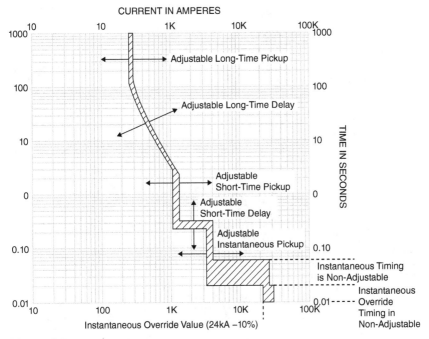

Figure 12.7 Protective Device Setting Basics.

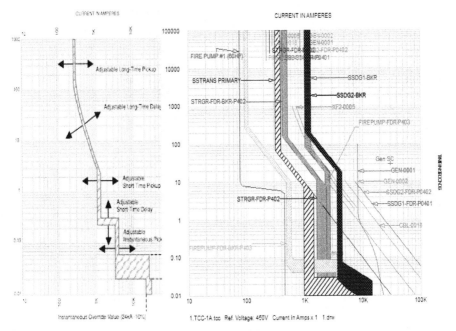

Figure 12.8 Protective Device Coordination.

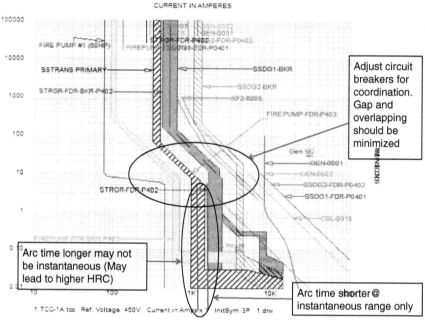

Figure 12.9   Protective Device Coordination-Typical.

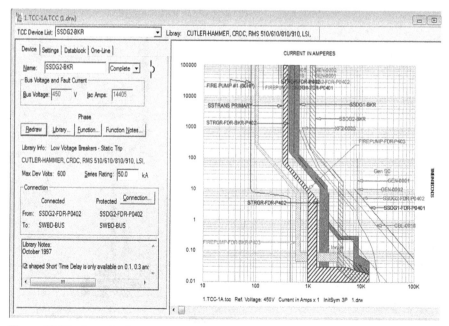

Figure 12.10   Protective Device Setting Options-1.

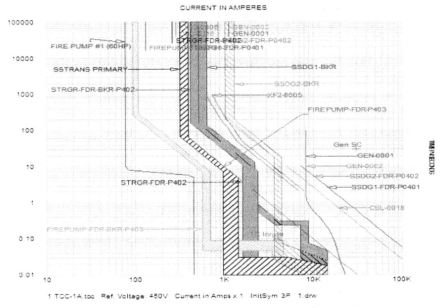

Figure 12.11   Protective Device Setting Options-2.

Table 12.9   Arc Flash Calculation Analysis and Spread Sheet Deliverables

Bus Name	Protective Device Name	Bus kV	Bus Bolted Fault (kA)	Bus Arcing Fault (kA)	Prot Dev Bolted Fault (kA)	Prot Dev Arcing Fault (kA)	Trip/ Delay Time (sec.)	Breaker Opening Time (sec.)	Duration of Arc (sec.)	Arc Type	Arc Flash Boundary (in)	Working Distance (in)	Incident Energy (cal/cm2)	Required Protective FR Clothing Category	Label #	Cable Length From Trip Device (ft)	Incident Energy at Low Marginal	Incident Energy at High Marginal
1 BUS-0021	SSTRANS PRIMARY	0.12	15.43	15.43	15.43	15.43	0.03	0.000	0.03	In Box	18	18	1.2	Category 0	# 0001	100.00		1.20
2 STEERING GEAR PANEL	STRGR-FDR-BKR-P402	0.45	11.20	11.20	11.20	11.20	0.02	0.000	0.02	In Box	11	18	0.59	Category 0	# 0002	100.00		
3 SWBD-BUS	SSDG2-BKR	0.45	14.40	5.47	7.14	2.71	2	0.000	2	In Box	118	18	19	Category 3 (*N3) (*N9)	# 0003			
4 Category 0: Nonmelting, Flammable Materials with Weight >= 4.5 oz/sq yd	0.0 - 1.2 cal/cm^2										#Cat 0 = 2		(*N3) - Arcing Current Low Tolerances Used					
5 Category 1: Arc-rated FR Shirt & Pants	1.2 - 4.0 cal/cm^2										#Cat 1 = 0		(*N9) - Max Arcing Duration Reached					
6 Category 2: Arc-rated FR Shirt & Pants	4.0 - 8.0 cal/cm^2										#Cat 2 = 0							
7 Category 3: Arc-rated FR Shirt & Pants & Arc Flash Suit	8.0 - 25.0 cal/cm^2										#Cat 3 = 1							
8 Category 4: Arc-rated FR Shirt & Pants & Arc Flash Suit	25.0 - 40.0 cal/cm^2										#Cat 4 = 0							
9 Category Dangerous: No FR Category Found	40.0 - 999.0 cal/cm^2										#Danger = 0		NFPA 70E-2009 Bus Report (80% Cleared Fault Threshold, include Ind. Motors for 5.0 Cycles), mis-coordination not checked					

12.22 ARC FLASH CALCULATION ANALYSIS AND SPREADSHEET DELIVERABLES

12.22.1 For Shipboard Arc Flash Analysis the Following Should Be Included

(a) Correct and complete electrical one-line diagram.

(b) Calculate bolted fault current at each bus.

(c) Calculate arcing fault current at each bus.

(d) Calculate arcing fault current seen by each protective device.

(e) Determine trip time for each protective device based on arcing fault current.

(f) Calculate incident energy at working distance.

(g) Calculate arc flash boundary.

(h) Determine PPE requirements.

(i) Generate labels.

(j) Train and retrain the operators periodically related to electrical safety matter in the ship. Periodicity as required or as agreed upon.

12.23 METHODS OF DEVELOPING ANALYSIS

The fault current analysis, coordination study, and arc flash study should be performed together. Since the arc flash study is a recent requirement, legacy fault current analysis-generated short-circuit ratings may not be suitable for the arc flash study.

12.23.1 Coordination Study

– Coordination is usually for systematic and coordinated equipment protection as well as system reliability.

– Arc flash requirements bring a new safety focus to the coordination study looking at minimum faults and then try to set a faster trip time. However, if the coordination is not done properly, the system may be under nuisance trip creating additional operational issues

12.24 FAULT CURRENT ANALYSIS TO ENSURE POWER SYSTEM COMPONENT PROTECTION CHARACTERISTICS

The system level fault current (overcurrent) protection is provided by using circuit breakers, contactors, and fuses. The circuit breaker is a switching mechanism capable of making, carrying, and breaking normal current for a specific time. The circuit breaker is also capable of breaking current under abnormal conditions without causing damage to the breaker within its operating range. The circuit breaker must be able to switch repeatedly under various conditions. The following definitions are provided to provide a better understanding of fault current behavior:

(a) **Symmetric current:** A periodic alternating current in which point a period apart are equal and have opposite sign

(b) **Asymmetric current:** A combination of the symmetric current plus the direct current component of the current

(c) **I $_{Average}$:** The average of the maximum asymmetrical rms current of the three-phase system fault at V cycle measured in amps.

(d) **I $_{max}$:** The maximum value of short-circuit current at V cycle in the phase having the maximum asymmetrical current.

(e) **Symmetrical fault current:** The symmetric fault current is an rms current which is identified as I symmetrical. I symmetrical is the initial available symmetrical short-circuit current, determined at V cycle after the inception of the fault.

(f) **I $_{min}$:** The minimum available rms asymmetrical current at the point of application of each circuit breaker.

12.25 FAULT CURRENT CALCULATION: APPROXIMATION FOR ARC FLASH ANALYSIS

(a) **For a given generator sub-transient reactance (6600 V system)**
5 MVA, 0.8 PF, three-phase, 60 Hz, 438 FLA and sub-transient reactance 18% as an example.
The maximum symmetrical RMS fault contribution:
at time $t = 0$ from each generator set is considered as $((1/0.18) \times 438) = 2433$ AMP
at around 100 msec the steady state short circuit current is $(3 \times 438) = 1,314$ AMP

(b) **For a given generator sub-transient reactance (690 V system)**
5 MVA, 0.8 PF, three-phase, 60 Hz, 4183 FLA and sub-transient reactance 16% as an example.
The maximum symmetrical RMS fault contribution:
at time $t = 0$ from each Generator set is considered as $((1/0.16) \times 4183) = 26,143$ AMP
at around 100 msec the steady state short circuit current is $(3 \times 4183) = 12,414$ AMP

(c) **For a given generator sub-transient reactance (450 V system)**
3 MVA, 0.8 PF, three-phase, 60 Hz, 3849 FLA and sub-transient reactance 15% as an example.
The maximum symmetrical RMS fault contribution:

at time $t = 0$ from each generator set is considered as $((1/0.15) \times 3849) = 25,660$ AMP

at around 100 msec the steady state short-circuit current is $(3 \times 3849) = 11,547$ AMP

12.26 SHIPBOARD FAULT CURRENT CALCULATION GUIDELINES (PER USCG REQUIREMENTS)

USCG CFR 46 §111.52-3 Systems below 1500 kilowatts.

The following short-circuit assumptions must be made for a system with an aggregate generating capacity below 1500 kilowatts, unless detailed computations in accordance with §111.52-5 are submitted:

(a) *The maximum short-circuit current of a direct current system must be assumed to be 10 times the aggregate normal rated generator currents plus six times the aggregate normal rated currents of all motors that may be in operation.*

(b) *The maximum asymmetrical short-circuit current for an alternating current system must be assumed to be 10 times the aggregate normal rated generator currents plus four times the aggregate normal rated currents of all motors that may be in operation.*

(c) *The average asymmetrical short circuit current for an alternating-current system must be assumed to be 81/2 times the aggregate normal rated generator currents plus 31/2 times the aggregate normal rated currents of all motors that may be in operation.*

12.27 EXAMPLE SHIPBOARD FAULT CURRENT CALCULATIONS (PER USCG REQUIREMENTS CFR 111-52-3(B) & (C))

The following fault current calculation is performed for a maximum of two generators (600 kW each) operating in parallel and with electric plant motor load which is assumed to be 900 kW.

The short-circuit current is calculated as follows:

(a) **Generator full load current:**

Each generator full load current

$= (700 \times 1000) / (\sqrt{3} \times 450 \times 0.8) = 1120$ AMP

(b) **Total motor contribution in amps:**

Motor running current $= (900 \times 1000) / (\sqrt{3} \times 450 \times 0.8 \times 0.93) = 1550$ AMP

(This assumes that the motors are approximately 93 % efficient and are designed for 0.8 power factor).

(c) Maximum fault current:

The maximum asymmetrical short-circuit current (2 generators in parallel) $= = 2 \times 10 \times 1120 + 6 \times 1550$

$\text{I}_{Max} = 22,400 + 9300 = 31,700 \text{ AMP}$

(d) Maximum asymmetrical fault current:

The maximum asymmetrical short-circuit current (2 generators in parallel) $= = 2 \times 10 \times 1120 + 4 \times 1550$

$\text{I}_{Max \ Assym} = 22,400 + 6200 = 28,600 \text{ AMP}$

(e) Average asymmetrical fault current:

The average asymmetrical current is $(2 \times 8.5 \times 1220 + 3.5 \times 1550)$

$\text{I}_{av} = 20,740 + 5425 = 26,165 \text{ AMP}$

(f) Minimum asymmetrical fault current (interpolating between maximum and minimum)

$\text{I}_{Min} = 24,940 \text{ AMP}$

12.28 SHIPBOARD POWER SYSTEM SHORT-CIRCUIT CURRENT CALCULATION (REFER TO US NAVY DESIGN DATA SHEET 300-2 FOR DETAILS)

$\text{I}_{Average}$ is the average of the maximum asymmetrical rms current of the three-phase system fault at $\frac{1}{2}$ cycle measured in amps. The equation is

$$I_{AVE} = \frac{0.71Eg}{\sqrt{\left(\frac{0.7Eg}{Im} + R_c\right)^2 + \left(\frac{0.19Eg}{Im} + X_c\right)^2}} + \frac{K_1Eg}{\sqrt{R^2 + X^2}}$$

$$I_{max} = \frac{Eg}{\sqrt{\left(\frac{0.7Eg}{Im} + R_c\right)^2 + \left(\frac{0.19Eg}{Im} + X_c\right)^2}}$$

$$I_{SYM} = \frac{0.63Eg}{\sqrt{\left(\frac{0.7Eg}{Im} + R_c\right)^2 + \left(\frac{0.19Eg}{Im} + X_c\right)^2}} + \frac{Eg}{\sqrt{R^2 + X^2}}$$

$$I_{MIN} = \frac{EG}{\sqrt{R^2 + X^2}}$$

$$I_{Min} = \frac{Eg}{\sqrt{R2 + X2}}$$

$$I_{Max} = \frac{K2Eg}{\sqrt{R^2 + X^2}}$$

$$I_{MIN} = \frac{Eg}{\sqrt{R^2 + X^2}}$$

Explanation of the terms

K1 the ratio of the average asymmetrical rms current in the three phases at one-half cycle to the rms value of the symmetrical current

K2 the ratio of the maximum rms asymmetrical current in one phase at one-half cycle to the rms value of the symmetrical current

E_g the line to neutral voltage of the system. For 480 V, three-phase system Eg is 480 divided by square root of 3

I_m the motor current contribution. For preliminary calculation Im is estimated two-thirds of the total connected generator ampere capacity

R the resistance of per phase circuit in ohms, looking from the fault to the generator

X the reactance per phase in ohms looking from the fault to the generator

Rc the per phase circuit resistance in ohms from the switchboard to the point of fault

Xc the per phase circuit reactance in ohms from the switchboard to the point of fault

Short-circuit calculation related decremental factors K1 and K2 values K1 = ratio of the average asymmetrical rms current in the three phases at one-half cycle to the rms value of the symmetrical current.

K2 = ratio of the maximum rms asymmetrical current in one phase at one-half cycle to the rms value of the symmetrical current.

For value of x/r less than 1.0, K1 = 1.00 and K2 = 1.00.

X/R ratio values between those listed are to be obtained by linear interpolation.

12.29 FAULT CURRENT AND ARC FLASH ANALYSIS AS REQUIRED BY NFPA 70E

NFPA 70E 2-13.3: "Flash hazard analysis shall be done before a person approaches any exposed electrical conductor or circuit part that has not been placed in an electrically safe work condition."

(Follow NPA-70E and National Electrical code guidelines and any other requirements to make shipboard electrical systems safe.)

12.30 FAULT CURRENT AND ARC FLASH ANALYSIS GUIDE BY IEEE 1584

Table 12.10   Fault Current Decrement Conversion Factors (K1 & K2) (from IEEE 45)

X/R	K1	K2	X/R	K1	K2
1.0	1.001	1.002	11.0	1.241	1.459
1.2	1.003	1.005	11.2	1.244	1.463
1.4	1.006	1.011	11.4	1.246	1.467
1.6	1.010	1.020	11.6	1.248	1.471
1.8	1.015	1.030	11.8	1.250	1.475
2.0	1.021	1.042	12.0	1.252	1.478
2.2	1.028	1.056	12.2	1.254	1.482
2.4	1.036	1.070	12.4	1.255	1.485
2.6	1.043	1.086	12.6	1.257	1.488
2.8	1.051	1.101	12.8	1.259	1.491
3.0	1.059	1.116	13.0	1.261	1.494
3.2	1.067	1.132	13.2	1.262	1.498
3.4	1.075	1.147	13.4	1.264	1.500
3.6	1.082	1.162	13.6	1.266	1.503
3.8	1.090	1.176	13.8	1.267	1.506
4.0	1.097	1.190	14.0	1.269	1.509
4.2	1.104	1.203	14.2	1.270	1.512
4.4	1.111	1.216	14.4	1.272	1.514
4.6	1.118	1.229	14.6	1.273	1.517
4.8	1.124	1.241	14.8	1.274	1.519
5.0	1.130	1.253	15.0	1.276	1.522
5.2	1.136	1.264	15.2	1.277	1.524
5.4	1.142	1.275	15.4	1.278	1.526
5.6	1.147	1.285	15.6	1.280	1.529
5.8	1.153	1.295	15.8	1.281	1.531
6.0	1.158	1.305	16.0	1.282	1.533
6.2	1.163	1.314	16.2	1.283	1.535
6.4	1.167	1.323	16.4	1.284	1.537
6.6	1.172	1.331	16.6	1.286	1.539
6.8	1.176	1.330	16.8	1.287	1.541
7.0	1.181	1.347	17.0	1.288	1.543
7.2	1.185	1.335	17.2	1.289	1.545
7.4	1.189	1.362	17.4	1.290	1.547
7.6	1.192	1.369	17.6	1.291	1.549
7.8	1.196	1.376	17.8	1.292	1.551
8.0	1.200	1.383	18.0	1.293	1.553
8.2	1.203	1.389	18.2	1.294	1.555
8.4	1.206	1.395	18.4	1.295	1.556
8.6	1.210	1.401	18.6	1.296	1.558
8.8	1.213	1.407	18.8	1.296	1.559
9.0	1.216	1.412	19.0	1.297	1.560
9.2	1.219	1.418	19.2	1.298	1.562
9.4	1.222	1.423	19.4	1.298	1.564
9.6	1.224	1.428	19.6	1.299	1.566
9.8	1.227	1.433	19.8	1.300	1.567
10.0	1.230	1.438	20.0	1.301	1.568
10.2	1.232	1.442	25.0	1.318	1.598
10.4	1.235	1.447	30.0	1.336	1.630
10.6	1.237	1.447	35.0	1.338	1.634
10.8	1.239	1.455	40.0	1.341	1.642

ELECTRICAL WORKPLACE SAFETY PRACTICE:

1. Find all possible sources of power supply to the equipment. Open the disconnecting device for each source.
2. Where possible visually verify the device is open.
3. Test voltage on each conductor to verify that it is really de-energized
4. Apply lock-out and tag-out devices.
5. De-energize equipment versus "work it live" unless increase hazard exist or infeasible due to design or operational limitations. Arc flash requirements bring new safety focus to coordination study looking at minimum faults and then trying to set faster trip time. However, if the coordination is not done properly, the system may be under nuisance trip, creating additional operational issues.
6. Switch remotely (if possible).
7. Close and tighten door latches or door bolts before operating a switch.
8. Stand to the side and away as much as possible during switching operations.

12.31 ELECTRICAL SAFETY AND ARC FLASH LABELING (NFPA 70E)

Arc Flash Labeling (Section 130.5(c)): Electrical equipment such as switchboards, panel boards, control panels, and motor controllers, motor control centers adjustment, servicing, or maintenance while energized, shall be field marked with a label containing the following information: (as applicable)

a. At least one of the following:

Available incident energy and the corresponding working distance
Minimum arc rating of the clothing
Required label PPE
Highest Hazard/Risk Category (HRC) for the equipment
Nominal system voltage
Arc flash boundary

b. New Annex-H table H.3(a) Provides guidance on application NFPA-70E sections related to shock hazard PPE and Arc flash PPE

c. New Table H.4(a) and Table H.4(b): Provide maximum three-phase bolted fault current limits at low voltage systems (208 V through 690 V) and high-voltage systems (5 kV, 12 kV, and 15 kV respectively), along with circuit breaker fault clearing times for the recommended use of 8 cal/cm^2 and 40 cal/cm^2 PPE in what is term an "arc -in-a-box" situation. The limitations noted in these tables are based on IEEE 1584 calculations methods.

110-8B-1.b: Arc Flash Hazard Analysis—An arc Flash Hazard analysis shall determine the ARC Flash Protection boundary and the personal protection equipment that people within the Arc Flash Protection Boundary shall use.

The best means to mitigate personnel safety concern for medium voltage power systems is to ensure the power system design has as many inherent personnel safety design aspects as practical. The following power system design features are recommended for medium-power systems:

- Well-trained personnel for maintenance and operation of medium voltage power system, including wearing proper PPE equipment, observing FPB safety boundaries, and using proper instrumentation and equipment
- Arc fault detection/protection system within medium-voltage switchboards and transformers
- Medium-voltage equipment, including electrical propulsion equipment, located in dedicated spaces that can be secured to limit personnel access or at least use only qualified personnel within that space and provide an arc flash area around medium-voltage equipment (such as facing arc flash vents in switchboards toward bulkheads, overheads or equipment enclosures, hull shell or sturdy sides of other equipment)
- Label all medium-voltage spaces, equipment, and cableways or cables as dangerous
- Color-code cabling (red, orange or yellow accepted as the industry standard to highlight medium-voltage hazard) for medium-voltage cables to distinguish them from other control, communication, and power cables
- Install medium-voltage cabling in separate cableways as remote from personnel access as practical
- Conduct rigorous inspection of all medium-voltage equipment installations before lite-off of power system
- Conduct periodic rigorous inspections of medium-voltage equipment

12.32 ARC FLASH PROTECTION-BOUNDARY

Table 12.11   Arc Flash Protection-Boundary

	Prohibited	Restricted	Limited	Remarks
480 VAC	1 inch (25 mm)	1 ft 0 inch (305 mm)	3 ft 6 inch (1.07 m)	
600 VAC	1 inch (25 mm)	1 ft 0 inch (305 mm)	3 ft 6 inch (1.07 m)	
4160 VAC	7 inch (187 mm)	2 ft 2 inch (660 mm)	5 ft 0 inch (1.52 m)	
6600 VAC	7 inch (187 mm)	2 ft 2 inch (660 mm)	5 ft 0 inch (1.52 m)	
13,800 VAC	7 inch (187 mm)	2 ft 2 inch (660 mm)	5 ft 0 inch (1.52 m)	

12.33 SAMPLE ARC FLASH CALCULATIONS: SPREADSHEET–EXCEL TYPE

The following methods are presented for arc flash calculations: (1) spreadsheet method, (2) NFPA 70E method, (3) IEEE 1584 method, and (4) advanced software method.

12.33.1 NFPA 70E 2009 Equation D.5.2 (A) for Arc Flash Calculation

Incident Energy Calculation

E_{MB} = Incident Energy in cal/cm² Maximum 20 cubic inch box

E_{MB} = 1 038.7 DB $^{-1.478}$ t $_A$ {0.0093F² − 0.3453F + 9.675}

D_B = Distance from arc electrodes in inches (For distances of 18 inches and more)

t_A = Arc duration in seconds

F = Short-circuit current in KA (for range between 16 kA to 50 kA)

12.34 LOW-VOLTAGE (50 V–1000 V) PROTECTION (NFPA 70E 130.3 (A1))

Voltage Levels Between 50 V and 1000 V: In those cases where detailed arc flash hazard analysis calculations are not performed for systems that are between 50 V and 1000 V, the **Arc Flash Protection Boundary shall be 4.0 ft,** based on the product of clearing time of 2 cycles (0.033 sec) and the available bolted fault current of 50 kA or any combination not exceeding 100 kA cycles (1667 ampere second).

When the product of clearing times and bolted fault current exceeds 100 kA cycles, the Arc Flash Protection Boundary shall be calculated.

12.35 MEDIUM VOLTAGE (1000 V AND ABOVE) (NFPA 70E 130.3 (A2))

At voltage levels above 1000 V, the Arc Flash Boundary shall be the distance at which the incident energy equals 5 J/cm² (1.2 cal/cm²). For situations where fault-clearing time is equal to or less than 0.1 sec, the Arc Flash Protection Boundary shall be the distance at which the incident energy level equals 6.24 J/cm² (1,5 cal/cm²).

12.36 IEEE 1584-BASED ARC FLASH CALCULATIONS

$Log_{10}E_n = K + K_2 + 1.081 * Log_{10} I_a + 0.0011 * G.$

E_n – Incident energy J/cm^2 normalized for time and distance. The equation above is based on data normalized for a distance from the possible arc point to the person of 610 mm (24 in), and an arcing time of 0.2 sec (12 cycle).

$K_1 = -0.792$ for open configurations, and is -0.555 for box configurations / enclosed equipment. (Use box configuration for shipboard application.)

$K_2 = 0$ for ungrounded and high resistance grounded systems, and equals -0.113 for grounded systems. (Use ungrounded and HRG system for shipboard application.)

G – Gap between conductors in millimeters.

I_a – Predicted three-phase arcing current in kA. It is found by using Equation 2 a) or b) so the operating time for protective devices can be determined.

12.36.1 IEEE 1584: Incident Energy Exposure

This is the amount of thermal incident energy to which the worker's face and chest could be exposed at working distance during an electrical arc5event. Incident energy is measured in joules per centimeter squared (J/cm^2) or in calories/cm^2 (5 J/cm^2 = 1.2 cal/cm^2). Incident energy is calculated using variables such as available fault current, system voltage, expected arcing fault duration and the worker's distance from the arc. The data obtained from the calculations is used to select the appropriate flame resistant (FR) PPE.

12.36.2 IEEE 1584: Arcing Current Calculation: Up to 1000 V Systems

$log_{10}I_a = K + 0.662 * log_{10} I\, bf + 0.0966 * V + 0.000526 * G + 0.5588 * V * log_{10}I\, bf - 0.00304 * G * log_{10}I\, bf$

... ... equation

log_{10} – is logarithm base 10 ($log_{10)}$.

I_a – arcing current in kA.

E_n – normalized incident energy in J/cm^2.

K – equals -0.153 for open configurations and -0.097 for box configurations. (Use box configuration for shipboard application.)

I_{bf} – bolted fault current for three-phase faults in kA symmetrical rms.

V – System voltage in kV.

G – gap between conductors in millimeters.

Solve $logI_a = 0.00402 + 0.983 * log\, I_{bf.}$

12.36.3 IEEE 1584: Arcing Current Calculation for 1 kV TO 15 kV

E – Incident energy exposure in J/cm^2.

$E = 4.184 * C_f * E_n * (t / 0.2) * (610^x/D^x)$

C_f – calculation factor equal to 1.0 for voltages above 1 kV, and 1.5 for voltages below 1 kV.

E_n – normalized incident energy in J/cm^2 as calculated above.

t – Arcing time in seconds.

D – Distance from possible arcing point to the person in millimeters.

x – Distance exponent.

12.36.4 IEEE 1584: Flash Protection Boundary Calculation (D_B)

D_B – distance of the boundary from the arc point in millimeters.

$D_B = [4.184 * Cf * En * (t / 0.2) * (610^x/EB)]^{1/x}$.

C_f – calculation factor equal to 1.0 for voltages above 1 kV, and 1.5 for voltages below 1 kV.

E_n – normalized incident energy in J/cm^2 as calculated.

E_B – incident energy in J/cm^2 at the boundary distance.

I_{bf} – bolted fault current for three phase faults in kA symmetrical rms.

t – arcing time in seconds.

x – Distance exponent.

E_B (incident energy in J/cm^2 at the boundary distance) is usually set at 5 J/cm^2 (1.2 cal/cm^2) for bare skin, or at the rating of proposed personal protection equipment.

12.36.5 IEEE 1584: Flash Protection Boundary

The flash protection boundary is an approach limit at a distance from exposed live parts or enclosed live parts if operation, manipulation, or testing of equipment creates a potential flash hazard, within which a person could receive a second degree burn if an electrical arc flash were to occur. A worker entering the flash protection boundary must be qualified and must be wearing appropriate PPE. The Flash Protection Boundary is required to be calculated by NFPA 70E.

12.36.6 IEEE 1584: Level of PPE

This is the minimum level of Personal Protective Equipment in calories per centimeter squared with the intent to protect the worker from the thermal effects of the arc flash at 18 inches from the source of the arc.

12.36.7 IEEE-1584: Equipment Class

Classes of equipment included in IEEE 1584 and typical bus gaps are shown in Table 12.12:

Table 12.12   IEEE 1584 Working Distance

Classes of Equipment	Typical Bus Gaps, mm
Open Air	10–40
Low-voltage switchgear	32
15 kV switchgear	152
5 kV switchgear	104
Low-voltage MCCs and panelboards	25
Cable	13

Typical working distance is the sum of the distance between the worker standing in front of the equipment, and from the front of the equipment to the potential arc source inside the equipment.

Arc-flash protection is always based on the incident energy level on the person's face and body at the working distance, not the incident energy on the hands or arms. The degree of injury in a burn depends on the percentage of a person's skin that is burned. The head and body are a large percentage of total skin surface area and injury to these areas is much more life threatening than burns on the extremities. Typical working distances are shown in Table 12.13:

Table 12.13

Classes of Equipment	Typical Working Distance, mm (in)
Low-voltage switchgear	610 mm (24 in)
15 kV / 5 kV switchgear	910
Low-voltage MCCs and panelboards	455
Cable	455

12.36.8 IEEE 1584: Distance Exponent

Table 12.14

System Voltage, kV	Equipment Type	Distance Exponent
	Open Air	2.0
0.208 - 1	Switchgear	1.473
	MCC and panels	1.641
	Cable	2.0
	Open Air	2.0
>1 to 15	Switchgear	0.973
	Cable	2.0

12.36.9 IEEE 1584: Arc Duration/Total Arc Clearing Time

Use protective device characteristics, which can be found in manufacturer's data. For fuses, the manufacturer's time-current curves may include both melting and clearing time. If so, use the clearing time. If they show only the average melt time, add to that time 15%, up to 0.03 seconds, and 10% above 0.03 seconds to determine total clearing time. If the arcing fault current is above the total clearing time at the bottom of the curve (0.01 seconds), use 0.01 seconds for the time.

For circuit breakers with integral trip units, the manufacturer's time-current curves include both tripping time and clearing time.

For relay operated circuit breakers, the relay curves show only the relay operating time in the time-delay region. For relays operating in their instantaneous region, allow 16 milliseconds on 60 Hz systems for operation. The circuit breaker opening time must be added. Opening times for particular circuit breakers can be verified by consulting the manufacturer's literature.

12.36.10 IEEE 1584: Available Three-Phase Bolted Fault Current

Available 3-phase bolted fault current for the range of 700 A to 106 kA at the point where work is to be performed in kA.

12.36.11 IEEE 1584: Predicted Three-Phase Arcing Current

The arcing current depends on the available 3-phase bolted fault current at the point where work is to be performed.

⚠ WARNING

ARC FLASH AND SHOCK HAZARD
APPROPRIATE PPE IS REQUIRED

ARC FLASH PROTECTION		SHOCK PROTECTION	
WORKING DISTANCE	18 INCHES	SHOCK HAZARD WHEN COVER IS REMOVED	
INCIDENT ENERGY – CALCULATED	8.0 CAL/CM²	VOLTAGE RATING	600 VAC
PPE RATING	LEVEL-2	LIMITED APPROACH	42 INCHES
		RESTRICTED APPROACH	12 INCHES
ARC FLASH PROTECTION BOUNDARY 48 INCH		PROHIBITED APPROACH	1 INCH

COMPANY'S SAFETY PROGRAM FOR PPE REQUIREMENTS

EQUIPMENT NAME & SYMBOL NUMBER: _____	ARC FLASH ANALYSIS BY: _____
LOCTION: _____	FILE # _____
	DATE: _____
	NFPA-70E-2012 AND IEEE-1580 COMPLIANT

NAME OF THE SHIP _____

Figure 12.12 Recommended AC Flash Warning for Ship.

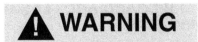

SHIPBOARD ARC FLASH HAZARD AND SHOCK IS PRESENT
CALCULATED HAZARD/RISK CATEGORY (HRC)
APPROPRIATE (PPE) IS REQUIRED

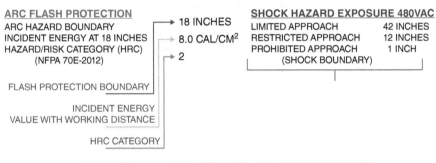

ARC FLASH PROTECTION
ARC HAZARD BOUNDARY → 18 INCHES
INCIDENT ENERGY AT 18 INCHES → 8.0 CAL/CM^2
HAZARD/RISK CATEGORY (HRC)
(NFPA 70E-2012) → 2

FLASH PROTECTION BOUNDARY

INCIDENT ENERGY
VALUE WITH WORKING DISTANCE

HRC CATEGORY

SHOCK HAZARD EXPOSURE 480VAC
LIMITED APPROACH   42 INCHES
RESTRICTED APPROACH   12 INCHES
PROHIBITED APPROACH   1 INCH
(SHOCK BOUNDARY)

HAZARD CATEGORY-2	HAZARD/RISK CATEGORY 2 (HRC-2) DESCRIPTION	
	COTTON UNDERWEAR	ARC RATED
	COTTON LONG SLEEVE SHIRT	ARC RATED
	SAFETY GLASS	
	INSULTED GLOVES	

Figure 12.13 Explanaton of the Arc Flash Sign.

HAZARD OF ELECTRIC SHOCK, EXPLOSION OR ARC FLASH

(1) Apply appropriate personal protective equipment (PPE) and follow safe electrical work practices per NFPA 70E
(2) This equipment must be installed and serviced by qualified electrical personnel
(3) Turn of all power supplying this equipment before working on or inside the equipment
(4) Use appropriate voltage sensing device to confirm power is off
(5) Use appropriate lock-out and tag out prior to working on the equipment
(6) Avail necessary work permit to work on the equipment

(7) Do not work alone in this equipment

Failure to follow these instructions will result in injury or death

Figure 12.14 Arc Flash Danger Sign.

12.37 SAMPLE SHIPBOARD ARC FLASH CALCULATION PROJECT

12.37.1 General

Before performing the Arc Flash Hazard Study, perform a short-circuit study for the equipment shown on the one-ine diagram. Compare the calculated short-circuit current to the equipment short-circuit ratings. Verify that the equipment is properly rated for the available short-circuit current. Protective device coordination study for the equipment to determine settings of the protective devices.

12.37.2 Short-Circuit Study

Perform a short-circuit study.

12.37.3 Protective Device Coordination Study

Perform a protective device coordination study for the equipment shown on the one line diagram. Determine the proper settings for the protective devices. For complex protective devices, create time current curves to determine the appropriate protective device sizes and settings.

12.37.4 Arc Flash Hazard Study

Perform the arc flash hazard study after the short-circuit and protective device coordination study have been completed.

Produce an Arc Flash Warning label. Labels shall be printed in color and shall be moisture proof, adhesive backed. Labels for outdoor equipment shall be vinyl and UV resistant to avoid fading.

Produce three Arc Flash Evaluation Summary Sheet reports. The first shall list the items below for the maximum calculated energy levels. The second report shall list all incident energy values at each location for each scenario. The energy level is 40 Cals/cm^2 (Above Hazard Risk Category 4).

12.37.5 Analysis

– Analyze the short-circuit, protective device coordination, and arc flash calculations, and highlight any equipment that is determined to be underrated or causes an abnormally high incident energy calculation.

– Propose general methods and approaches to reduce energy levels. Proposed major corrective modifications will be taken under advisement by the engineer, and the contractor will be given further instructions.

12.37.6 Report

The results of the shipboard power system arc flash analysis shall be summarized in a final report. The report shall include the following sections:

Introduction, executive summary and recommendations, assumptions, and a reduced copy of the one-line drawings.

Arc Flash Evaluations Summary Spreadsheet (Maximum Energy Calculation)

Arc Flash Hazard Warning Labels printed in color on paper. One set of labels printed in color on moisture proof adhesive backed labels. Labels for outdoor equipment shall be vinyl and UV resistant to avoid fading.

12.38 FAST-ACTING ARC MANAGEMENT SYSTEM: ARC FLASH MITIGATING HARDWARE DRIVEN TIME LIMITING FEATURES:

Traditional arc protection methods with manipulating protective device settings may not provide fast enough protection as short arc burning time is critical, especially when the arc develops during maintenance work on the switchgear, endangering personnel safety and life. The unique fast-acting arc fault functionality adds a new dimension to the total safety of the installation and the reliability of the power system protection. This effective means of limiting arc flash hazards is accomplished by limiting the arcing time using a dedicated arc flash detection relay and fiber optical sensor. Arc flash detecting relays typically reduce the arc flash incident energy by nearly instantaneous detection of the arc flash using very fast reacting sensors, such as light sensors. This provides faster tripping of upstream breakers for arc flash faults, minimizing the arcing time and thereby reducing the incident energy level in the fault.

Key Features of fast-acting arc detection system:

- Flexible and modular system can be adapted to different targets requiring arc protection
- Central unit and modular units engineer a scheme to your requirements
- Continuous system self-supervision
- Operation on simultaneous current and light or on light only
- Direct connection of arc sensors in the central unit without using I/O units
- Typically 7 millisec operation time with trip contact and 2 millisec with high speed output

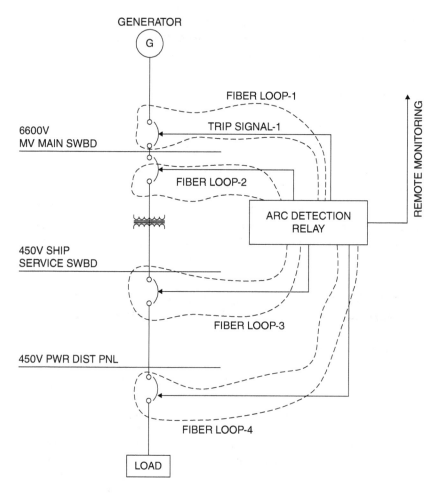

NOTE-1: THE FAST ACTING ARC DETECTION RELAY PICKS UP FIBER OPTICAL SIGNAL WHICH CAN INITIATE CIRCUIT BREAKER TRIP SIGNAL IN ABOUT 2.5MSEC INITIATING APPROPRIATE CIRCUIT BREAKER TRIP. THIS FAST ACTION REDUCES THE ARCHING ENERGY.

Figure 12.15   Optical Sensor Schematics for Arc Protection System.

- Programmable operation zones
- Communication protocol support for SCADA and automation interfacing

12.39 GUIDELINES FOR SHIPBOARD PERSONNEL

a. Always consider the circuit is live unless it is established that the circuit is dead. Therefore "do not start," "do not open," "do not energize," and "do not operate."

b. Follow safety procedure of "Lockout," and "Tag-out."

(In some cases, the work to be carried out is not subject to a safety procedure and doesn't need a permit to work into which a lock-out tag-out procedure is incorporated. In this case, one can decide to protect oneself with a lockout tag-out procedure.)
Sequential Procedure:

Preparation for shutdown

Machine or equipment shutdown

Machine or equipment isolation

Lockout or tag-out device application

Stored energy

Verification of isolation

Preparation for Shutdown: Before turning off a machine or equipment one has to have knowledge of:

• Type and magnitude of the energy.
• The hazards of the energy to be controlled
• Method or means to control the energy Machine or Equipment Shutdown
• The machine or equipment must be turned off or shut down using the procedures established, to avoid any additional or increased hazards to employees as a result of the machine or equipment stoppage Machine or equipment isolation
• All energy-isolating devices that are needed to control the machine's energy source must be located

From the figure, we can see that if the fault happens on the "ESWBD" (Critical Switchboard), the arcing current of 15.95 kA going through, "CSWB MCB" will clear the fault at 0.215 seconds. This will produce a total incident energy of 11.3 cal/cm^2.

Now, if the instantaneous setting of the "CSWB MCB" device is changed from 20 to 10, it will then clear the fault at 0.018 seconds. This will then produce a total reduced incident energy of 1.0 cal/cm^2.

Category-1: Arc Flash Technical Guide Subcategory: Mitigation

An example of how changing the STPU of a protective could also reduce the incident can be seen in Figures 3 and 4. Figure 3 shows a partial single line modeled in SKM software along with its corresponding time current curve (TCC). From Figure 3, we can see that if the fault happens on bus "MCC#1A," the arcing current of 9.38 kA going through, "52-SUB3A-MCC1A" will clear the fault at 0.24 seconds. This will produce a total incident energy of about 10.4 cal/cm^2.

Category-2: Arc Flash Technical Guide Subcategory: Mitigation

Now, if he STPU setting of the "52-SUB3A-MCC1A" device is reduced from 0.2 to 0.1, as in Figure 4, it will then clear the fault at 0.15 seconds. This will then produce an arc-flash incident energy of 6.7 cal/cm^2.

Note that special care must be taken when changing protective devices to mitigate arc flash. One should not just haphazardly lower the settings to reduce the tripping time of devices. One must be careful that no overlapping of TCC or mis-coordination is mistakenly achieved on other parts of the system when the settings are lowered. Mis-coordination could cause nuisance tripping or even increase the incident energy on the other parts of the system.

Category-3: Arc Flash Technical Guide Subcategory: Mitigation

Another effective way of lowering the arc flash incident energy is by temporarily overriding the breaker's or relay's delay function to trip without intentional delay whenever a fault is detected. This can be achieved by applying a maintenance switch or multiple settings group during the maintenance mode of operation.

An ARMS (Arc Resistance Maintenance Switch) is a device that can be retrofitted with certain existing trip units, such that, when the ARMS is switched on, the tripping time of the unit is very fast when a fault is detected. When a person wants to perform maintenance, the maintenance switch is turned on. The breaker's delay functions are automatically overridden and the breaker then trips instantaneously if a fault is detected. When a maintenance task is completed, the switch is turned off and all previous trip unit settings are reactivated.

A multiple settings group works in a similar fashion as the ARMS. Here, you configure two relays in series, such that when you turn a switch on "maintenance" you have two curves on your TCC. One curve is for the normal operation and the other curve is an instantaneous curve such that when there's a fault, the relay sends a signal to trip really fast.

The next three figures illustrate how the ARMS device works, modeled in the SKM software. Figure 5 shows a partial single line modeled in SKM software along with its corresponding time current curve (TCC).

From the figure, we can see that if the fault happens on the line side "52-CRITICAL-MN," the arcing current of 16.02 kA going through, "52-SUB2A-UPS#1" will clear the fault at 0.216 seconds. This will produce a total incident energy of 11.7 cal/cm^2.

Glossary

ABS American Bureau of Shipping.

ACC Automatic Centralized Control. This notation is assigned by the ABS to a vessel having the means to control and monitor the propulsion-machinery space from a continuously manned centralized control and monitoring station installed within or adjacent to the propulsion machinery space.

ACCU Automatic Centralized Control Unmanned. This notation is assigned by the ABS to a vessel having the means to control andmonitor the propulsion-machinery space from the navigation bridge and from a centralized control and monitoring station installed within or adjacent to the propulsion machinery space.

ALARM Visual and audible signals indicating an abnormal condition of a monitored parameter (ABS).

ALARM Alarm means an audible and visual indication of a hazardous or potentially hazardous condition that requires attention (USCG).

AMBIENT TEMPERATURE Ambient temperature is the temperature of surrounding media such as air or fluid where equipment is operated or positioned.

ARC FLASH A type of electrical arcing fault, electrical explosion, or electrical discharge that results from a low impedance electric current path through air to a conductive plan or to another voltage phase in an electrical system.

ASD Adjustable speed drive. ASD can be a cycloconverter type, load commutating inverter type, pulse width modulation type, etc.

AUTOMATED Automated means the use of automatic or remote control, instrumentation or alarms (USCG) (4-9-1/5.1.4).

AUTOMATIC CONTROL means of control that conveys predetermined orders without action by an operator (ABS).

AUTOMATIC CONTROL Automatic control means self regulating in attaining or carrying out an operator-specified equipment response or sequences (USCG).

AUXILIARY SERVICES SYSTEM All support systems (e.g., fuel oil system, lubricating system, cooling water system, compressed air system, hydraulic system, etc.) that are required to run propulsion machinery and propulsors (ABS).

AZIMUTH THRUSTER Rotatable mounting thruster device where the thrust can be directed to any desirable direction.

BANDWIDTH Generally, frequency range of system input over which the system will respond satisfactory to a command.

BLACKOUT RECOVERY

BRAKING Braking provides a means of stopping and can be accomplished by dynamic braking, regenerative braking, DC injection braking, and positive action brake.

BREAKDOWN TORQUE The maximum torque that can be developed with rated parameters, such as rated voltage applied at rated frequency.

Shipboard Power Systems Design and Verification Fundamentals, First Edition. Mohammed M. Islam.
© 2018 the Institute of Electrical and Electronics Engineers, Inc. Published 2018 by John Wiley & Sons, Inc.

CENTRALIZED CONTROL STATION A propulsion control station fitted with instrumentation, control systems, and actuators to enable propulsion and auxiliary machinery be controlled and monitored, and the state of propulsion machinery space be monitored, without the need of regular local attendance in the propulsion machinery space (ABS).

CIRCUIT BREAKER FRAME (1) The circuit breaker housing that contains the current carrying components, the current sensing components, and the tripping and operating mechanism. (2) That portion of an interchangeable trip molded case circuit breaker remaining when the interchangeable trip unit is removed (100 AF, 400 AF, 800 AF, 1600 AF, etc.).

Continuous current rating (ampere rating) The designated RMS alternating or direct current in amperes which a device or assembly will carry continuously in free air without tripping or exceeding temperature limits.

Drawout circuit breaker An assembly of a circuit breaker and a supporting structure (cradle) so constructed that the circuit breaker is supported and can be moved to either the main circuit connected or disconnected position without removing connections or mounting supports.

Electronic trip circuit breaker A circuit breaker that uses current sensors and electronic circuitry to sense, measure, and respond to current levels.

IDMT (Inverse Definite Minimum Time) Time/current graded overcurrent protection. Basically, the more current put through the relay, the faster it goes.

Instantaneous pickup The current level at which the circuit breaker will trip with no intentional time delay.

Instantaneous trip A qualifying term indicating that no delay is purposely introduced in the tripping action of the circuit breaker during short-circuit conditions.

Insulated case circuit breaker (ICCB) UL Standard 489 Listed non-fused molded case circuit breakers that utilize a two-step stored energy closing mechanism, electronic trip system, and drawout construction.

Interrupting rating The highest current at rated voltage available at the incoming terminals of the circuit breaker. When the circuit breaker can be used at more than one voltage, the interrupting rating will be shown on the circuit breaker for each voltage level. The interrupting rating of a circuit breaker must be equal to or greater than the available short-circuit current at the point at which the circuit breaker is applied to the system.

Inverse time A qualifying term indicating there is purposely introduced a delay in the tripping action of the circuit breaker, which delay decreases as the magnitude of the current increases.

KAIC Kilo Amperes interrupting capacity. (65 KAIC or 65,000 AIC)

Long-time ampere rating An adjustment that, in combination with the installed rating plug, establishes the continuous current rating of a full-function electronic trip circuit breaker.

Long-time delay The length of time the circuit breaker will carry a sustained overcurrent (greater than the long-time pickup) before initiating a trip signal.

Long-time pickup The current level at which the circuit breaker long-time delay function begins timing.

Making capacity (of a switching device) The value of prospective making current that a switching device is capable of making at stated voltage-prescribed conditions of use and behavior.

Molded case circuit breaker (MCCB) A circuit breaker assembled as an integral unit in a supportive and enclosed housing of insulating material, generally 20 to 3000 A in size and used in systems up to 600 VAC and 500 VDC.

Peak let-through current The maximum peak current flowing in a circuit during an overcurrent condition.

Short-circuit delay (STD) The length of time the circuit breaker will carry a short circuit (current greater than the short-circuit pickup) before initiating a trip signal.

Short-circuit making capacity Making capacity for which prescribed conditions include a short circuit at the terminals of the switching device.

Short-circuit breaking capacity Breaking capacity for which prescribed conditions include a short circuit at the terminals of the switching device

Time current curve (TCC) Method of ensuring selective coordination is to examine each overcurrent device's time-current curve (TCC) and verify for any value of current, that the protective device closest to the fault clears faster than any upstream device.

CLEARING TIME The total time between the beginning of the overcurrent and the final opening of the circuit at rated voltage by an overcurrent protective device (ABS 4-9-1/5.1.15).

CONSTANT HORSEPOWER RANGE A range of motor operation where motor speed is greater than base rating of the motor, in the case of AC motor operation usually above 60 Hz where the voltage remains constant as the frequency is increased.

CONSTANT TORQUE RANGE A speed range in which the motor is capable of delivering a constant torque, subject to motor thermal characteristics. This essentially is when the inverter/motor combination is operating at constant volts/Hz.

CONSTANT VOLTS/HERTZ (V/Hz) This relationship exist in AC drives where the output voltage is varied directly proportional to frequency. This type of operation is required to allow the motor produce constant rated torque as speed is varied.

CONTINUOUS RATED MACHINE The continuous rating of a rotating electrical machine is the rated kW load at which the machine can continuously operate without exceeding the steady state temperature rise (ABS) (4-9-1/5.1.2).

CONTROL The process of conveying a command or order to enable the desired action be effected (ABS) (-1/5.1.3).

CONTROL SYSTEM An assembly of devices interconnected or otherwise coordinated to convey the command or order.

CONTROLLABLE PITCH PROPELLER Propeller pitch is changeable, mostly hydraulic.

CONVERTER The process of changing AC to DC, or AC to DC to AC. A machine or device for changing AC power into DC power (rectifier operation) or DC power into AC power (inverter operation).

Current-Source Converter A current source converter is characterized by a controlled DC current in the intermediate DC link. The line side network voltage is converted in a controlled DC current. A current source converter always uses an SCR bridge or an active front end to control the DC current.

Voltage-Source Converter A voltage source converter is characterized by a stiff DC voltage in the intermediate DC link. The line side network voltage is converted to a constant DC voltage. A voltage source converter uses a diode bridge (6-pulse, 12-pulse, 18-pulse, 24-pulse or 36-pulse) or an active front end to connect the DC link to the network voltage.

CURRENT LIMITING An electronic method of limiting the maximum current available to the motor. This is adjusted so that the motor's maximum current can be controlled. It can also be preset as a protective device to protect both the motor and control from extended overloads.

DEVIATION Difference between an instantaneous value of a controlled variable and the desired value of the controlled variable corresponding to the set point.

DMM Digital multi-meter.

DOL starter Direct online starting system.

DRIVE A drive designed to provide easily operable means for speed adjustment of the motor within a specified speed range. The equipment used for converting electrical power into mechanical power suitable for the operation of a machine. A drive is a combination of a converter, motor, and any motor mounted auxiliary devices. Examples of motor mounted auxiliary devices are encoders, tachometers, thermal switches and detectors, air blowers, heaters, and vibration sensors.

DUTY CYCLE The relationship between the operating and rest times or repeatable operation at different loads.

EFFICIENCY Ratio of mechanical output to electrical input indicated by percent.

ENGINEERING CONTROL CENTER (ECC) ECC means a centralized engineering control, monitoring, and communications location (USCG).

ENCLOSURE Enclosure refers to the housing in which the equipment is mounted (ABS 4-9-1/5.1.11).

FAIL-SAFE A designed failure state that has the least critical consequence. A system or a machine is fail-safe when, upon the failure of a component or subsystem or its functions, the system or the machine automatically reverts to a designed state of least critical consequence (ABS).

FAIL-SAFE Fail-safe means that upon failure or malfunction of a component or subsystem, the output automatically reverts to a predetermined design state of least critical consequence (USCG).

FAILURE MODE AND EFFECT ANALYSIS (FMEA) A failure analysis methodology used during design to postulate every failure mode and the corresponding effect or consequences. Generally, the analysis is to begin by selecting the lowest level of interest (part, circuit, or module level). The various failure modes that can occur for each item at this level are identified and enumerated. The effect for each failure mode, taken singly and in turn, is to be interpreted as a failure mode for the next higher functional level. Successive interpretations will result in the identification of the effect at the highest function level, or the final consequence. A tabular format is normally used to record the results of such a study (ABS).

FIXED PITCH PROPELLER The propeller blades are fixed. There is no possibility of changing propeller pitch.

FULL-LOAD TORQUE The full-load torque of a motor is the torque necessary to produce rated horsepower at full-load frequency.

INDUCTION MOTOR An alternating current motor in which the primary winding on one member (usually stator) is connected to the power source. A secondary winding on the rotor carries the induced current (ABS 4-9-1/5.1.5).

INSTRUMENTATION A system designed to measure and display the state of a monitored parameter and which may include one of more of sensors, read-outs, displays, alarms, and means of signal transmission (ABS).

INVERTER A converter in which the direction of power flow is predominately from the DC terminal to the AC terminal. – [B13]

> **Current Source Inverter (CSI)** An inverter in which the DC terminal is inductive and, as a consequence, the DC current is relatively slow to change. Modulation of the CSI acts to control the voltage at the AC terminal. The switches in a CSI must block either voltage polarity, but are only required to conduct current in one direction.

> **Voltage Source Inverter (VSI)** An inverter in which the DC terminal is capacitive and, as a consequence, the DC voltage is relatively stiff. Modulation of the VSI acts to control the current at the AC terminal. The switches in a VSI must block DC voltage, but be able to conduct current in either direction.

IPDE Integrated Product Data Environment, which features the capability to concurrently develop, update, and reuse data in electronic form.

IPT Integrated product team composed of representatives from appropriate disciplines working together to build successful programs, identify and resolve issues, and make sound and timely recommendations to facilitate decision-making.

ISOCHRONOUS Constant speed and frequency irrespective of load.

LCI Type of adjustable speed current source drive called load commutating inverter.

LOCAL CONTROL A device or array of devices located on or adjacent to a machine to enable it be operated within sight of the operator (ABS).

LOCAL CONTROL Local control means operator control from a location where the equipment and its output can be directly manipulated or observed, e.g., at the switchboard, motor controller, propulsion engine, or other equipment (USCG).

LOCKED ROTOR TORQUE The minimum torque of a motor that will develop at rest for all angular positions of the rotor, with rated voltage applied at rated frequency.

LO/TO Lockout and tag-out.

MANUAL CONTROL Manual control means operation by direct or power-assisted operator intervention (USCG) (4-9-1/5.1.9).

MONITORING SYSTEM A system designed to supervise the operational status of machinery or systems by means of instrumentation, which provides displays of operational parameters and alarms indicating abnormal operating conditions (ABS).

MULTIVIBRATORS Sequential logic circuits is used to build complex circuits such as multivibrators, counters, shift registers, latches, and memories, etc, These circuits operate in a "sequential" way, and require a clock pulse or timing signal to cause them to change their state. Clock pulses are generally a continuous square or rectangular shaped waveform that is produced by a single pulse generator circuit such as a multivibrator.

NFPA National Fire Protection Association.

NIBS Navigational Integrated Bridge System.

NON-PERIODIC DUTY RATED MACHINE The non-periodic duty rating of a rotating electrical machine is the kW loads which the machine can operate continuously, for a specific period of time, or intermittently under the designed variations of the load and speed within the permissible operating range, respectively; and the temperature rise, measured when the machine has been run until it reaches a steady temperature condition, is not to exceed those given in 4-8-3/Table 4 (ABS).

OPENING TIME (of mechanical switching device) Interval of time between the specified instant of initiation of the opening operation and the instant when the arcing contacts have separated in all poles. For circuit breakers: For a circuit breaker operating directly, the instant of initiation of the opening operation means the instant when the current increases to a degree big enough to cause the breaker to operate.

OSHA Occupational Safety and Health Administration.

OVERCURRENT A condition that exists on an electrical circuit when the normal full load current is exceeded. The overcurrent conditions are overloads and short circuits.

PERIODIC DUTY RATING MACHINE The periodic duty rating of a rotating machine is the rated kW load at which the machine can operate repeatedly, for specified period (N) at the rated load followed by a specified period (R) of rest and de-energized state, without exceeding the temperature rise given in 4-8-3/Table 4; where $N + R = 10$ min, and the cyclic duty factor is given by $N/(N + R)$ % (ABS) (4-8).

PODDED PROPULSION A propulsion electric motor is installed in a watertight enclosure, where the motor is mounted on the same motor shaft.

PPE Personal protective equipment.

PROPULSION MACHINE A device (e.g., diesel engine, turbine, electric motor, etc.) that develops mechanical energy to drive a propulsor (ABS).

PROPULSION MACHINERY SPACE Any space containing machinery or equipment forming part of the propulsion system (ABS).

PROPULSION SYSTEM A system designed to provide thrust to a vessel consisting of: one or more propulsion machines; one or more propulsors; all necessary auxiliaries, and associated control, alarm and safety systems (ABS).

PROPULSOR A device (e.g., propeller, or waterjet) that imparts force to a column of water in order to propel a vessel, together with any equipment necessary to transmit the power from the propulsion machinery to the device (e.g., shafting, gearing etc.) (ABS).

PROTECTION DEVICES

Breaking current Value of prospective breaking current that a device is capable of breaking at a stated voltage under prescribed conditions of use and behavior.

Breaking time Interval of time between the beginning of the opening time of a mechanical switching device and the end of the arcing time.

Circuit breaker frame (1) The circuit breaker housing that contains the current carrying components, the current sensing components, and the tripping and operating mechanism. (2) That portion of an interchangeable trip molded case circuit breaker remaining when the interchangeable trip unit is removed (100AF, 400AF, 800AF, 1600AF etc.).

Continuous current rating (ampere rating) The designated RMS alternating or direct current in amperes which a device or assembly will carry continuously in free air without tripping or exceeding temperature limits.

Drawout circuit breaker An assembly of a circuit breaker and a supporting structure (cradle) so constructed that the circuit breaker is supported and can be moved to either the main circuit connected or disconnected position without removing connections or mounting supports.

Electronic trip circuit breaker A circuit breaker that uses current sensors and electronic circuitry to sense, measure, and respond to current levels.

IDMT: Inverse Definite Minimum Time Time/current graded overcurrent protection. Basically, the more current put through the relay, the faster it goes.

Instantaneous pickup The current level at which the circuit breaker will trip with no intentional time delay.

Instantaneous trip A qualifying term indicating that no delay is purposely introduced in the tripping action of the circuit breaker during short-circuit conditions.

Insulated case circuit breaker (ICCB) UL Standard 489 Listed non-fused molded case circuit breakers that utilize a two-step stored energy closing mechanism, electronic trip system, and drawout construction.

Interrupting rating The highest current at rated voltage available at the incoming terminals of the circuit breaker. When the circuit breaker can be used at more than one voltage, the interrupting rating will be shown on the circuit breaker for each voltage level. The interrupting rating of a circuit breaker must be equal to or greater than the available short-circuit current at the point at which the circuit breaker is applied to the system.

Inverse time A qualifying term indicating there is purposely introduced a delay in the tripping action of the circuit breaker, which delay decreases as the magnitude of the current increases.

KAIC Kilo Amperes interrupting capacity (65 KAIC or 65,000 AIC).

Long-time ampere rating An adjustment that, in combination with the installed rating plug, establishes the continuous current rating of a full-function electronic trip circuit breaker.

Long-time delay The length of time the circuit breaker will carry a sustained overcurrent (greater than the long-time pickup) before initiating a trip signal.

Long-time pickup The current level at which the circuit breaker long-time delay function begins timing.

Making capacity (of a switching device) Value of prospective making current that a switching device is capable of making at a stated voltage under prescribed conditions of use and behavior.

Molded case circuit breaker (MCCB) A circuit breaker assembled as an integral unit in a supportive and enclosed housing of insulating material, generally 20 to 3000 A in size and used in systems up to 600 VAC and 500 VDC.

Opening time (of mechanical switching device) Interval of time between the specified instant of initiation of the opening operation and the instant when the arcing contacts have separated in all poles. For circuit breakers: For a circuit breaker operating directly, the instant of initiation of the opening operation means the instant when the current increases to a degree big enough to cause the breaker to operate.

Peak let-through current The maximum peak current flowing in a circuit during an overcurrent condition.

Short-circuit delay (STD)   The length of time the circuit breaker will carry a short circuit (current greater than the short-circuit pickup) before initiating a trip signal.

Short-circuit making capacity   Making capacity for which prescribed conditions include a short circuit at the terminals of the switching device.

Short-circuit breaking capacity   Breaking capacity for which prescribed conditions include a short circuit at the terminals of the switching device

Time Current Curve (TCC)   Method of ensuring selective coordination is to examine each overcurrent device's time-current curve (TCC) and verify for any value of current, that the protective device closest to the fault clears faster than any upstream device.

PWM   PWM is a DC inverter system with the main power being rectified to produce DC and a self-commutated inverter to invert DC to AC to a variable frequency.

R1   This is ABS optional notation assigned to a vessel fitted with multiple propulsion machines but only a single propulsor and steering.

R2   This is ABS optional notation assigned to a vessel fitted with multiple propulsion machines and also multiple propulsors and steering system.

R1-S   This is ABS optional notation assigned to a vessel fitted with only single propulsor but having the propulsion machinery arranged in separate spaces such that a fire or flood in one space will not effect the propulsion machinery in the other space.

R2-S   This is an optional notation assigned to a vessel fitted with multiple machines and propulsors, and associated steering systems arrangement in separate spaces such that a fire or flood in one space will not effect the propulsion machine(s) and propulsor(s), and associated steering systems in the other space.

R1, R2, R1-S and R2-S (Plus)(+)   This notation signifies that the redundant propulsion is capable of maintaining position under adverse weather conditions to avoid uncontrolled drift.

REDUNDANT SYSTEM:

REDUNDANCY DEFINITION BY ABS:

Redundancy Design Autonomous System: An autonomous system is a system that can control and operate itself, independently of any control system or auxiliary systems not directly connected to it.

Closed Bus: Closed bus often describes an operational configuration where all or most sections and all or most switchboards are connected together, that is, the bus-tie breakers between switchboards are closed. The alternative to closed bus is open bus, sometimes called split bus or split ring. Closed bus is also called joined bus, tied bus or closed-ring.

Common Mode Failure: A common mode failure occurs when events are not statistically independent, when one event causes multiple systems to fail. Critical Redundancy: Equipment provided to support the worst case failure design intent.

Differentiation: Differentiation is a method to avoid common mode failures by introducing a change of producer of redundant systems based on the same principle.

Fail Safe Condition: The system is to return to a safe state in the case of a failure or malfunction.

Hidden Failure: a failure that is not immediately evident to operations and maintenance personnel, such as protective functions on which redundancy depends.

Independence: An independent system is a system that can operate without the assistance of central control or other systems or subsystems. In this Guide it is mainly in reference to main machinery such as generators and thrusters. Auxiliary and control functions are to be provided in a manner that makes the machinery as independent as practical to minimize the number of failures that can lead to the loss of more than one main item of machinery.

Noncritical Redundancy: Equipment provided over and above that required to support the worst case failure design intent. Its purpose is to improve the reliability and availability of systems.

Redundancy: Ability of a component or system to maintain or restore its function, when a single fault has occurred. Redundancy can be achieved for instance by installation of multiple components, systems or alternative means of performing a function.

Redundancy Concept: The means by which the worst case failure design intent is achieved. It is to be documented as part of the preliminary design process.

Two or more component groups each of which is capable of individually and independently performing a specific function.

Separation: (Redundant systems) Separation is to reduce the number of connections between systems to reduce the risk that failure effects may propagate from one redundant system to another.

Single Fault: The single fault is an occurrence of the termination of the ability to perform a required function of a component or a subsystem.

REMOTE CONTROL A device or array of devices connected to a machine by mechanical, electrical, pneumatic, hydraulic or other means and by which the machine may be operated remote from, and not necessarily within sight of, the operator (ABS).

REMOTE CONTROL Remote control means nonlocal automatic or manual control (USCG) (ABS 4-9-1/5.1.8).

REMOTE CONTROL STATION A location fitted with means of remote control and monitoring (ABS).

RMS CURRENT The RMS is root-mean-square. The RMS current is the root-mean-square value of any periodic current.

SAFETY SYSTEM An automatic control system designed to automatically lead machinery being controlled to a predetermined less critical condition in response to a fault which may endanger the machinery or the safety of personnel and which may develop too fast to allow manual intervention. To protect an operating machine in the event of a detected fault, the automatic control system may be designed to automatically (ABS):

slowdown the machine or to reduce its demand;
start a standby support service so that the machine may resume normal operation; or
shutdown the machine.

For the purposes of this chapter, automatic shutdown, automatic slow down, and automatic start of standby pump are all safety system functions. Where "safety system" is stated herein after, it means any or all three automatic control systems.

SHORT TIME DUTY RATING MACHINE The short time rating of a rotating electrical machine is the rated kW load at which the machine can operate for a specified time period without exceeding the temperature rise given in 4-8-3/Table 4. A rest and de-energized period sufficient to reestablish the machine temperature to within 20 °C (3.60 °F) of the coolant prior to the next operation is to be allowed. At the beginning of the measurement the temperature of the machine is to be within 50 °C (90 °F) of the coolant (ABS).

SLIP The difference between the rotating magnetic field speed (synchronous speed) and rotor speed of an AC induction motor, which is usually expressed as a percentage of synchronous speed.

STEERING SYSTEM A system designed to control the direction of a vessel, including the rudder, steering gear, etc. (ABS) (4-9-1/5.1.12).

SYSTEMS INDEPENDENCE Systems are considered independent where they do not share components, such that a single failure in any one component in a system will not render the other systems inoperative (ABS).

THD Total harmonic distortion. It is accumulated distortion of fundamentals in a variable frequency drive.

THYRISTOR A component that conducts current in only one direction. Unlike a diode, the thyristor needs a firing pulse for it to start conducting current, after which it continues to conduct as long as there is current through it.

TOC Total ownership cost comprised of costs to research, develop, acquire, own, operate, maintain, and dispose of any and all assets comprising the ship.

TUNNEL THRUSTER Produces fixed directional thrust (ABS 4-9-1/5.1.14).

UNMANNED PROPULSION MACHINERY SPACE Propulsion machinery space that can be operated without continuous attendance by the crew locally in the machinery space and in the centralized control station (ABS 4-9-1/5.1.17).

UPS Uninterruptiable power system.

NO-BREAK POWER Power transfer from one source to another source instantaneously.

VITAL AUXILIARY PUMPS Vital auxiliary pumps are that directly related to and necessary for maintaining the operation of propulsion machinery. For diesel propulsion engines, the fuel oil pump, lubricating oil pump, and cooling water pumps are examples of vital auxiliary pumps (ABS).

VITAL SYSTEM OR EQUIPMENT Vital system or equipment is essential to the safety of the vessel, its passengers, and crew (USCG).

Index

ABS (American Bureau of Shipping), 3, 15–16, 23–24, 32, 37, 42, 45, 47–48, 104–106, 113, 118, 120, 122, 158, 177, 184–185, 212
ABS-SVR, 16, 23, 24, 32
ACP (alternate compliance plan), 185
AFE (active front end), 51, 62, 86, 88–89, 155, 160, 173, 179, 222, 267
AHF (active harmonic filter), 171–172, 179
ANSI (American National Standard Institute), 20, 24, 38, 55, 135–136, 140, 142–146, 153, 180, 187, 229, 287
arc flash, 2, 19, 20–21, 59, 65, 126, 180–184, 236–237, 239–240, 260, 262, 275–284, 286–293, 295, 297–298, 300, 302–311, 313–314
ASD (adjustable speed drive), 1–2, 53, 59, 61, 151–154, 159–160, 172, 177–179, 185–186, 211–212
AVR (automatic voltage regulator), 71, 103–104, 110–111, 177, 179, 261

BJT (bipolar junction transistor), 62–64, 161

cable termination, 38, 91, 171
CAMO (critical activity mode of operation), 25, 27–28, 267
CFR (circular of federal register), 4, 14, 19, 31, 74, 76–77, 104, 115–116, 136–137, 276, 286, 298
circuit breaker frame, 126
circuit breaker trip, 18, 58, 141, 145, 183, 312
common mode voltage, 211–212
CPU (central processing unit), 77
CSA (Canadian Standard Association), 46, 187, 192, 281
CSI (current source inverter), 153

D&D margin (design and development margin), 114
DF (distortion factor), 165
DG (diesel generator), 70–71
differential protection, 128–129, 136, 140, 142, 144–146, 156, 264, 270, 278
DVTP (design verification test procedure), 5, 7, 33, 66, 69, 73, 75, 77–78, 273

EMC (electromagnetic compatibility), 14, 36, 155, 196, 212
emergency generator, 4, 6, 16–18, 23–24, 33, 79, 81, 86, 88, 98, 100, 109–110, 114, 116–122, 136, 204, 261
EMI (electromagnetic interference), 30, 61–63, 91, 93, 160–161, 177, 192, 212
EOL (elementary one line), 4–7, 9, 132–133, 137, 139–143, 145–146, 264–265
EPA (Environmental Protection Agency), 49–50, 52–53
ETO (electrotechnical officer), 4
exciter transformer, 74

fault current calculations, 57–58, 131, 136, 138, 297–298
feeder, 14, 31, 38–39, 56, 69, 72, 74, 110, 117, 122, 125, 131, 134, 140, 145, 149–151, 153, 155–157, 186, 204, 215, 230, 233, 242, 246–248, 250, 270–271
filter, 14, 24, 27, 30–31, 36, 49–53, 61–66, 87–88, 91, 109, 155, 158–160, 162, 167–172, 175, 179, 181–182, 222, 228, 237–239, 267
FLA (full load ampere), 152, 154
FMEA (failure mode and effect analysis), 25–26, 28, 36, 67–68, 70, 73–74, 261, 264, 266–267, 272–273

Shipboard Power Systems Design and Verification Fundamentals, First Edition. Mohammed M. Islam.
© 2018 the Institute of Electrical and Electronics Engineers, Inc. Published 2018 by John Wiley & Sons, Inc.

CPSIA information can be obtained
at www.ICGtesting.com
Printed in the USA
LVHW050830180623
749271LV00004B/11